I0446638

DIY Off-Grid Solar Power for Everyone

Step by Step Guide to Design, Install, and Maintain Solar Systems for Homes, RVs, Vans, and Boats

After reading this book, please consider leaving an honest review.

It would be highly appreciated

Debasish Dutta

For more interesting projects, video tutorials, product reviews and updates, you can connect with me through the following platforms:

Blogpost: https://www.opengreenenergy.com

YouTube: https://www.youtube.com/c/opengreenenergy

Instagram: https://www.instagram.com/opengreenenergy

Instructables: https://www.instructables.com/member/opengreenenergy

Copyright

Copyright © 2023 Debasish Dutta all rights reserved.

The transmission, duplication, or reproduction of any of the following work, including specific information, will be considered an illegal act irrespective of whether it is done electronically or in print. This extends to creating a secondary or tertiary copy of the work or a recorded copy and is only allowed with the expressed written consent of the publisher. All additional rights reserved.

No part of this Book may be reproduced, distributed, or transmitted in any other form or by any other means, without the prior written permission of the publisher, except in the case of brief quotations embodied in critical reviews and certain other non-commercial uses permitted by copyright law.

For permission requests, write to the publisher at opengreenenergy@gmail.com

This Book is provided for informational purposes only and does not constitute professional advice. Although the author and publisher have made every effort to ensure the accuracy and completeness of the information contained in this book, they assume no responsibility for errors, inaccuracies, omissions, or any inconsistency herein. Readers should consult a professional before making any decisions based on the content of this Book.

Affiliation and Endorsement Disclaimer

Any product names, logos, brands, and other trademarks or images featured or referred to within this book are the property of their respective trademark holders. These trademark holders are not affiliated with the author, this book, or our website. They do not sponsor or endorse this book or any of our online products. The author declares no affiliation, sponsorship, nor any partnerships with any registered trademarks.

Preface

I have always been fascinated by using the sun's energy, something that has been around forever. As technology got better and we learned more about being green, using solar power became not only doable but also really good. And when it comes to using this endless energy, off-grid solar systems are a top choice for being independent and eco-friendly.

I have spent countless hours researching and delving into this topic, sharing my findings and insights through articles on my blog and Instructables. I have covered everything from basic concepts to more intricate projects like DIY MPPT Charge Controller. Over time, it became clear to me that having all this valuable information spread out over different posts was less than ideal. I felt it could be more beneficial if it was all in one place. That's why I decided to compile everything, leading to the creation of this book.

To access colour images and schematics accompanying this book, visit https://www.opengreenenergy.com/ebook01. This special page is exclusively for book owners, so remember to use the password found on the last page to gain access.

Intended audience

This book is for anyone who wants to use solar power, whether it's for an RV, van, boat, cabin, or tiny home. If you are new to solar or have some experience but need more guidance, this is the right place for you. I have made sure to cover all the basics and dive into some advanced stuff too. The goal is to help you understand and set up solar systems that fit your unique living space and needs.

Acknowledgments

Behind every endeavour, there are pillars of support, and I owe immense gratitude to a few:

My parents: Their unwavering faith in my potential and the values they instilled in me have been foundational. They taught me perseverance and the value of chasing one's passions, a lesson that fuels every page of this book.

My wife: Her relentless support has been the cornerstone of this journey. In moments of self-doubt or challenge, it was her faith and encouragement that reignited my passion, motivating me to press forward.

My parental aunt: Her impact on my life cannot be overstated. The vivid memories of her carrying me on her shoulders to my early morning tuitions exemplify her dedication to my education. Her consistent motivation and belief in the value of learning have been instrumental in shaping my journey.

My dear friend, Sajid Khan: His ability to transform complex concepts into lucid and engaging diagrams has truly elevated this work. His contribution has been pivotal in making the knowledge in this book accessible.

My social media community: The feedback, questions, and interactions I've had with you all have been nothing short of enlightening. Your engagement has ensured this book is relevant, answering real questions and catering to genuine curiosities.

Lastly, to my ever-growing community of readers and social media followers, your engagement has been the compass guiding this project. Every question, feedback, and discussion played a pivotal role in shaping the content, ensuring it meets the needs and curiosities of a diverse audience.

Dive in, soak up the knowledge, and let's journey together towards a brighter, more sustainable future powered by the sun.

Warm regards,

Debasish Dutta

Contents

CHAPTER 1 SOLAR BASICS AND FUNDAMENTALS   1
1.1 Introduction to Solar Energy   1
1.2 Solar Energy Conversion Process   1
1.3 How Solar Panels Are Made?   3
1.4 Solar Panel Manufacturing Process: 4

CHAPTER 2 INTRODUCTION TO OFF-GRID SOLAR SYSTEMS   7
2.1 What is an off-grid solar system?   7
2.2 Why Go Off-Grid?   9

CHAPTER 3 UNDERSTANDING ELECTRICITY BASICS   11
3.1 Basic Electrical Concepts: Voltage, Current, Resistance, and Power   11
3.2 Electricity Water Analogy   12
3.3 Understand Of Ohm's Law   13
3.4. Series and Parallel Connection   16
3.5 Difference Between AC and DC   18
3.6 Simple Math for Electrical Calculations   19
3.7 Energy Cost   21
Measuring Instruments   22
3.8 Digital Multimeter   22
3.9 Ammeter or Clamp Meter   26
3.10 Energy Meter   29

CHAPTER 4 SOLAR BATTERIES   31
4.1 Types of Lead Acid Battery   31
4.2 Types of Lithium Battery   37
4.3 Lithium Iron Phosphate (LiFePO4 or LFP) Batteries   38
4.4 Battery State of Charge (SoC)   42
4.5 Battery Depth of Discharge (DoD)   44
4.6 Battery Cycle Life   44
4.7 Battery C Rating   45
4.8 Battery Charging   48
4.9 Effect of Temperature on Battery   51
4.10 Series and Parallel Connection   52

CHAPTER 5 SOLAR PANELS   55
5.1 What is a Solar Panel?   55
5.2 Solar Spectrum Overview   56

5.3 Solar Irradiance   57

5.4 Radiation Resource Map   62

5.5 Types of Solar Panels   64

5.6 New Solar Panel Technologies in the Market   66

5.7 Emerging Solar Panel Technologies   68

5.8 Conversion Efficiency   69

5.9 Characteristics Curves   70

5.10 STC and NOCT   76

5.11 Ambient Temperature and Cell Temperature   78

5.12 Temperature coefficients   79

5.13 Series and Parallel Connections   82

5.14 Combining Different Solar Panel Types   84

5.15 Tilt & Azimuth Angle   85

5.16 Shading   88

5.17 Solar Panel Hotspot   89

5.18 Bypass Diodes and Blocking Diodes   91

CHAPTER 6 SOLAR CHARGE CONTROLLER   95

6.1 What is Role of a Charge Controller?   95

6.2 Types of Charge Controller   95

6.3 Charging Stages of Solar Charge Control   100

6.4 Charge Controller Protection Features   101

6.5 Choosing the Right Solar Charge Controller   101

6.6 Example Calculation   103

CHAPTER 7 OFF GRID SOLAR INVERTER   107

7.1 What is Role of an Inverter?   107

7.2 How Does An Off Grid Inverter Work?   108

7.3 Types of Off-grid Solar Inverter   109

7.4 Inverter / Charger   110

7.5 Inverter Power Rating   111

7.6 Inverter Efficiency   112

7.7 Key Features of Off-Grid Solar Inverters   113

CHAPTER 8 ESSENTIAL TOOLS FOR DIY SOLAR INSTALLATION   115

8.1 Screwdrivers   115

8.2 Cordless Impact Drill   116

8.3 Wire Strippers   117

8.4 Cable Strippers   118

8.5 Wire/Cable Cutter   119

8.6 Needle Nose Plier   120

8.7 Crimping tool   120

8.8 Torpedo Level   125

8.9 MC4 Spanners 125

8.10 Hex Nut Ratchet Sets 127

8.11 Safety Equipment 128

CHAPTER 9 EQUIPMENT USED IN OFF GRID SOLAR SYSTEM 131

9.1 Solar Panel Connectors 131

9.2 Crimp Connectors 137

9.3 Cable Lugs 139

9.4 Fuses 140

9.5 Circuit Breakers 145

9.6 DC Isolator Switch 149

9.7 DC Bus bars 150

9.8 Surge Protection Device (SPD) 152

9.9 Combiner Box 154

CHAPTER 10 EARTHING AND LIGHTNING PROTECTION 157

10.1 What is Earthing (Grounding)? 157

10.2 Why is Earthing Important? 157

10.3 Two Types of Grounding 158

10.4 Key Components to Ground in an Off-Grid Solar System 159

10.5 Typical Grounding Method for Mobile Setups 159

10.6 Lightning Protection 160

10.7 DIY or Professional Installation? 162

CHAPTER 11 SIZING OF OFF-GRID SOLAR COMPONENTS 163

11.1 Daily Energy Consumption 163

11.2 Solar Panel Sizing 163

11.3 Battery Bank Sizing 165

11.4 Sizing of Charge Controller 166

11.5 Sizing and Selecting an Inverter 167

11.5 Sizing Fuses and Circuit Breakers 168

CHAPTER 12 CABLE SELECTION AND SIZING 171

12.1 Types of Cables Used in Off-Grid Solar Systems 171

12.2 Factors to Consider When Selecting Cables 173

12.3 Understanding Wire Gauge Systems 175

12.4 Wire Gauge Table 176

12.5 Calculating Appropriate Cable Size 178

12.6 DC Cable Size Chart 180

CHAPTER 13 DIY OFF GRID SOLAR POWER INSTALLATION 181

13.1 Site Assessment 181

13.2 Planning and Design 182

13.3 Gathering Materials   183

13.4 Mounting Solar Panels   183

13.5 Battery Bank Setup   184

13.6 Installing Charge Controller   185

13.7 Inverter Installation   186

13.8 DC Fuse Box Installation   189

13.9 Wiring the Solar Panels   190

13.10 Final Testing   191

13.11 Example Wiring Diagrams   192

CHAPTER 14 MAINTENANCE AND CARE   195

14.1 Solar Panels   195

14.2 Battery Bank   197

14.3 Inverter and Charge Controller   198

14.4 Balance of System (BOS) Components   200

CHAPTER 15 COST AND FINANCING OPTIONS   203

15.1 Cost Factors for Off-Grid Solar System Installation   203

15.2 Financing Options   205

15.3 Evaluating Return on Investment (ROI)   206

15.4 Example Calculation   207

15.5 Example Calculation for ROI   209

CHAPTER 16 OFF-GRID SOLAR PERMIT   211

16.1 Permitting for Off-Grid Solar   211

16.2 Solar Permit Guide   212

16.3 FAQs About Off-Grid Solar Permitting   214

RECOMMENDED BRANDS   215
CONCLUSION   218

Chapter 1
Solar Basics and Fundamentals

1.1 Introduction to Solar Energy

Solar energy has been harnessed and used by humans for thousands of years. From the ancient Greeks and Romans using passive solar design in their architecture to more recent advancements in photovoltaic technology, solar energy has played a significant role in the development of sustainable and renewable energy solutions.

The importance of solar energy in the modern world cannot be overstated. As the global demand for energy continues to grow, the need for clean, renewable sources of power has become increasingly critical. Solar energy offers a virtually limitless supply of power that is both environmentally friendly and economically viable. By harnessing the power of the sun, we can reduce our dependence on fossil fuels, decrease greenhouse gas emissions, and promote energy independence.

Solar energy has a wide range of applications, from residential rooftop installations to utility-scale solar farms. In addition to generating electricity, solar energy can also be used for water heating, space heating, and even cooling in some cases. Furthermore, solar-powered devices, such as solar lanterns and portable chargers, have become increasingly popular, providing essential energy services in off-grid or remote areas.

The advancements in solar energy technology have made it more accessible and affordable than ever before. With continuous research and development, solar energy has the potential to become a significant contributor to our global energy mix, paving the way for a more sustainable and cleaner future. This chapter aims to introduce the fundamentals of solar energy, including the conversion process, and how solar panels are made.

The solar energy conversion process involves capturing sunlight and converting it into usable electricity. This is achieved through a phenomenon known as the photovoltaic (PV) effect. The PV effect is the process by which certain materials, called semiconductors, can generate an electric current when exposed to sunlight.

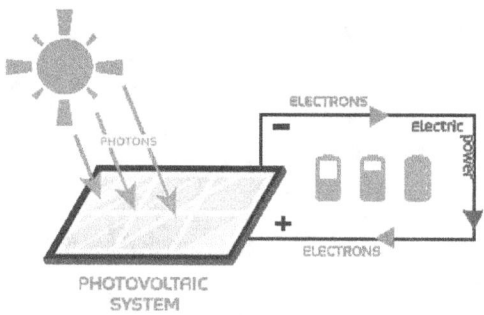

Fig 1.1

A solar cell is the fundamental unit responsible for this conversion process. The most common material used in solar cells is silicon, which is a semiconductor. A solar cell is typically composed of two layers of silicon – one layer is doped with impurities to create an excess of electrons (called an n-type layer), while the other layer is doped to create a deficit of electrons (called a p-type layer). When these two layers are brought together, an electric field is formed at the junction between them, creating a region called the depletion zone.

When sunlight, composed of particles called photons, hits the solar cell, some photons are absorbed by the semiconductor material. The energy from these absorbed photons is transferred to the electrons in the silicon atoms, causing them to break free from their atomic bonds. These free electrons are then pushed by the electric field towards the n-type layer, while the "holes" they leave behind are pushed towards the p-type layer.

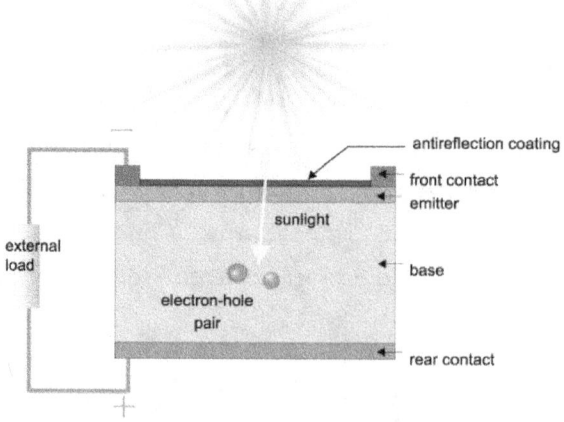

Fig 1.2

By attaching metal contacts to both layers of the solar cell, an external electrical circuit is created. When the free electrons flow through this circuit, an electric current is generated. This direct current (DC) can then be used to power electrical devices, or it can be converted into alternating current (AC) using an inverter, making it suitable for use in homes and businesses.

1.3 How Solar Panels Are Made?

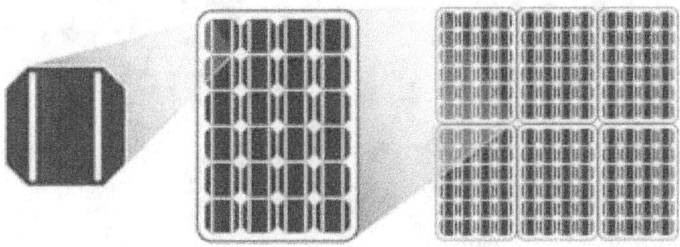

Fig 1.3

Solar panels are made up of multiple solar cells connected together in series or parallel configurations to produce a higher voltage or current output. The combined output of the solar cells in a panel forms the basis of a photovoltaic (PV) module. PV modules can then be connected together to form a solar array, capable of generating a significant amount of electricity to meet various energy needs.

In the next section, we will discuss the detailed process of manufacturing solar panels from raw silica.

1.4 Solar Panel Manufacturing Process:

Solar cells, commonly made from silicon, are the fundamental building blocks of a photovoltaic solar panel. They convert sunlight directly into electricity. Here's a basic outline of how traditional silicon-based solar cells are made:

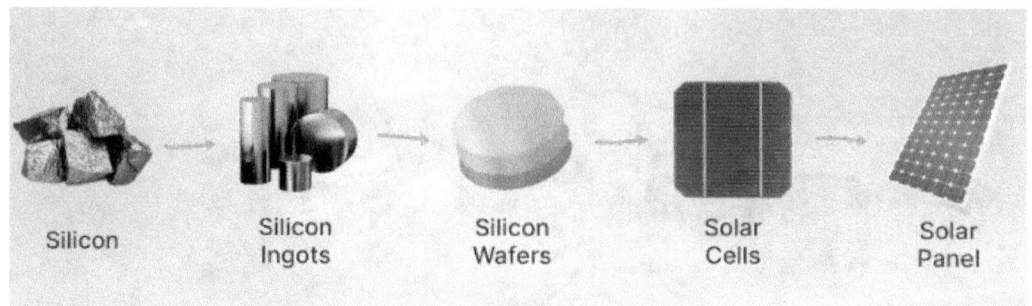

Fig 1.4

Image Credit: https://www.solarreviews.com/

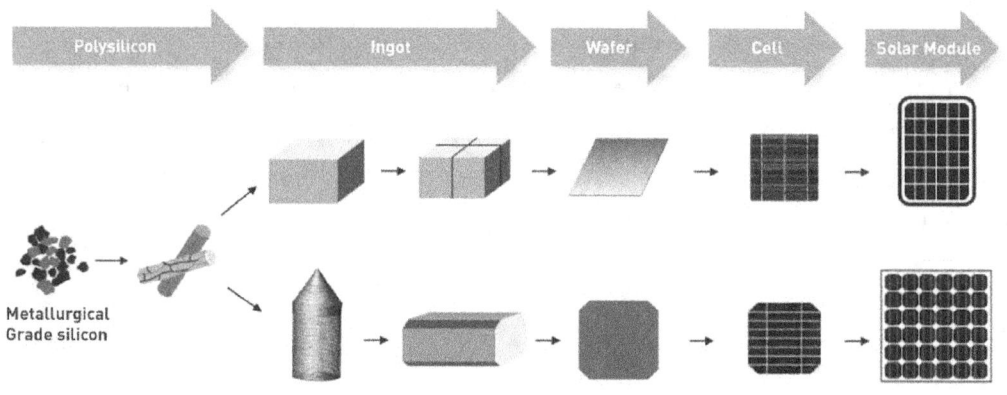

Fig 1.5

1. **Purifying Silicon:**

   Silicon is found in nature in the form of silicon dioxide (like some types of sand and quartz). To extract pure silicon, the silicon dioxide is heated in a reduction furnace with carbon.

   The result is metallurgical-grade silicon, which is further refined to electronics-grade silicon using multiple rounds of purification.

2. **Making Monocrystalline or Polycrystalline Silicon:**

   Monocrystalline: Silicon is melted and then slowly drawn out in the form of a single cylindrical crystal, known as a boule.

   Polycrystalline: Silicon is melted and then allowed to cool and solidify in a mold, creating multiple crystals.

3. **Slicing Silicon into Wafers:**

   The silicon block (boule or block) is sliced into very thin wafers using a diamond saw. These wafers serve as the substrate for the solar cells.

4. **Surface Texturing:**

   The wafers are textured to create a rough surface. This increases light absorption, thereby increasing the efficiency of the solar cell.

5. **N-Type and P-Type Layer Creation:**

   In order to generate an electric current, solar cells are constructed as PN junctions. This is usually achieved by introducing small amounts of different materials (doping) into the silicon wafer.

   For example, introducing boron will create a positive (P-type) layer, while introducing phosphorus will create a negative (N-type) layer. The junction of these layers is where electron movement will occur when exposed to sunlight.

6. **Anti-Reflective Coating:**

   An anti-reflective coating is applied to the silicon wafer to reduce the amount of light that's reflected off the surface. This helps increase the amount of light absorbed and, consequently, the efficiency of the solar cell.

7. **Electrical Contacts:**

   Metal contacts are deposited on the top and bottom of the solar cell to extract the current generated when sunlight hits the cell. Typically, the front side has a grid-like pattern to allow sunlight to still penetrate the cell, while the back is usually fully covered.

8. Assembly into Solar Panels:

Individual solar cells are then interconnected (usually in series) and assembled into larger panels. The interconnected cells are encapsulated between layers of protective materials (often glass on the front, and a polymer-based material on the back).

9. Final Module Protection and Framing:

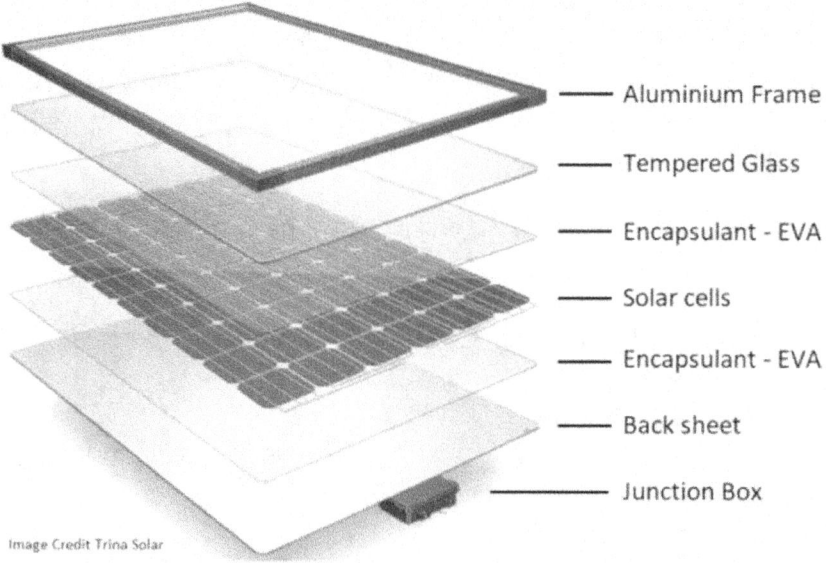

- Aluminium Frame
- Tempered Glass
- Encapsulant - EVA
- Solar cells
- Encapsulant - EVA
- Back sheet
- Junction Box

Image Credit Trina Solar

Fig 1.6

To protect solar cells, a durable glass layer is added, and an encapsulant usually made of ethylene-vinyl acetate (EVA) secures the cells in their proper position.

The back sheet offers insulation and contributes to the overall structural integrity of the panel. Meanwhile, an aluminium frame provides support and makes mounting the panel easier. The junction box houses essential electrical connections and bypass diodes, which prevent power loss due to shading

Chapter 2
Introduction to Off-Grid Solar Systems

2.1 What is an off-grid solar system?

An off-grid solar system, also known as a stand-alone power system (SAPS), is a self-contained electrical generation and storage system that is entirely independent of the utility grid. Off-grid solar systems generate electricity through solar panels and store the energy in batteries to be used as needed, day or night. They are typically used in remote locations, where connecting to the main electrical grid is impractical or too expensive, or by individuals who wish to achieve energy independence and reduce their environmental impact.

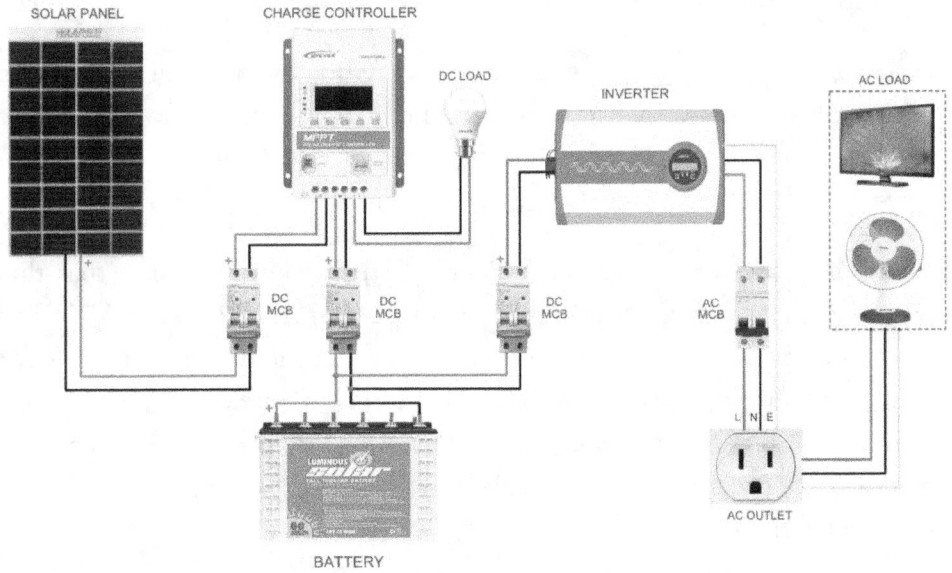

Fig 2.1

An off-grid solar system consists of several key components that work together to convert sunlight into usable electricity:

1. **Solar panels:** These are the primary energy source for the system. Solar panels, also known as photovoltaic (PV) panels, convert sunlight into direct current

(DC) electricity. The number and size of solar panels needed depend on the energy requirements of the location and the amount of sunlight available.

2. **Charge controller:** The charge controller regulates the flow of electricity from the solar panels to the battery bank. It ensures that the batteries are charged efficiently and prevents them from overcharging, which can lead to reduced battery life and potential safety hazards.

3. **Battery bank:** The battery bank stores the electricity generated by the solar panels for use when there is little or no sunlight, such as during cloudy days or at night. Batteries used in off-grid solar systems are typically deep-cycle batteries, designed to provide consistent energy output and withstand frequent charging and discharging.

4. **Inverter:** Most household appliances and electronics use alternating current (AC) electricity, while solar panels produce DC electricity. The inverter's role is to convert the DC electricity from the solar panels and batteries into AC electricity that can be used by standard appliances.

5. **Wiring and safety devices**: The system's wiring connects all the components, allowing electricity to flow between them. It is essential to use appropriate cables and safety devices, such as fuses and circuit breakers, to protect the system from electrical faults and potential hazards.

Off-grid solar systems require careful planning and design to ensure that they can provide enough electricity to meet the energy needs of the users. This involves calculating the required solar panel and battery capacity, selecting appropriate components, and designing a solar panel array that maximizes energy generation. Additionally, off-grid solar systems may require more significant attention to energy conservation and efficiency, as the available power supply is limited by the size of the solar array and battery bank.

Fig 2.2

There are several reasons why individuals and families choose to go off-grid and opt for a stand-alone power system. Some of the most common motivations include:

- **Remote location:** In many cases, off-grid solar systems are installed in remote or rural areas where connecting to the utility grid is not feasible or cost-effective. Extending power lines to these locations can be prohibitively expensive, making off-grid solar systems a more practical and economical solution.

- **Energy independence:** By generating and storing their own electricity, off-grid solar system users can become entirely self-sufficient and not reliant on utility companies for their power needs. This can offer a sense of freedom and control over energy production and consumption.

- **Environmental impact:** Off-grid solar systems harness clean, renewable energy from the sun, reducing the reliance on fossil fuels and decreasing greenhouse gas emissions. For environmentally conscious individuals, going off-grid can be an effective way to minimize their carbon footprint.

- **Reliability:** Off-grid solar systems are not subject to power outages and fluctuations that may affect the utility grid. By having an independent power source, users can maintain a consistent and reliable supply of electricity, even in the event of natural disasters or grid failures.

- **Cost savings:** While the initial investment in an off-grid solar system can be substantial, the long-term savings can outweigh the costs. Off-grid system users can avoid monthly utility bills and potential increases in electricity rates. Additionally, there are often financial incentives and tax credits available for installing renewable energy systems, further reducing the overall cost.

- **Scalability and customization:** Off-grid solar systems can be designed and scaled to meet the specific energy needs of the users. This allows for greater flexibility and customization compared to grid-tied systems, which may be subject to utility company restrictions and regulations.

- **Learning and personal growth:** Building and maintaining an off-grid solar system can be an educational and rewarding experience. By engaging in the process of designing, installing, and maintaining the system, users can develop a deeper understanding of energy production and management, as well as valuable skills in problem-solving, resourcefulness, and self-reliance.

Chapter 3
Understanding Electricity Basics

3.1 Basic Electrical Concepts: Voltage, Current, Resistance, and Power

To design, install, and maintain an off-grid solar system, it's essential to have a basic understanding of the fundamental concepts of electricity. This section will introduce the core principles of voltage, current, resistance, and power.

1. **Voltage (V):**

   Voltage, measured in volts (V), represents the electrical potential difference between two points in a circuit. It is the force that pushes electric charge (in the form of electrons) through a conductor, such as a wire. In a solar system, the voltage output of the solar panels and batteries determines the amount of electrical potential available to power appliances and devices.

2. **Current (I):**

   Current, measured in amperes or amps (A), represents the flow of electric charge through a conductor. It is analogous to the flow of water through a pipe; the greater the current, the more charge is being transferred. In a solar system, the current is directly related to the amount of power generated by the solar panels and the power consumed by the connected devices.

3. **Resistance (R):**

   Resistance, measured in ohms (Ω), is the opposition to the flow of electric current in a conductor or circuit. Resistance causes some of the electrical energy to be converted into heat, which can result in power loss. In a solar system, resistance can occur in the wiring, connectors, and various components. Minimizing resistance is crucial for maximizing the efficiency of the system.

4. **Power (P):**

   Power, measured in watts (W), represents the rate at which electrical energy is converted into other forms of energy, such as light or heat. It is the product of voltage and current ($P = V \times I$). In a solar system, power is generated by the solar panels, stored in the batteries, and consumed by the connected appliances and devices.

Understanding these basic electrical concepts is crucial for the successful design and operation of an off-grid solar system. In the following sections, we will explore units of measurement, simple math for electrical calculations, and how these concepts are applied in solar system design and sizing.

3.2 Electricity Water Analogy

The voltage, current, and resistance in an electrical circuit can be explained using the water analogy, which compares the flow of electricity to the flow of water through a pipe. This analogy helps to simplify and visualize the concepts for those who are new to electronics or have difficulty understanding these electrical parameters.

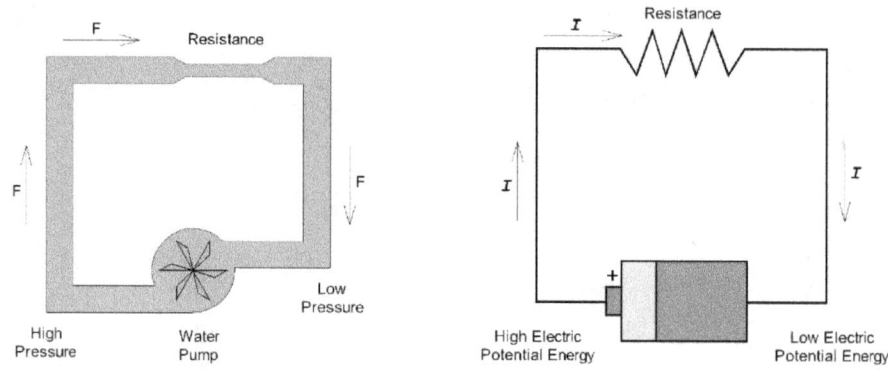

Fig 3.1

1. **Voltage (Water Pressure)**

   In the water analogy, voltage can be compared to water pressure. The greater the water pressure in a pipe, the more force it has to push water through the pipe. Similarly, voltage is the force or "electrical pressure" that pushes electric charge (measured in coulombs) through a conductor or circuit. Voltage is typically measured in volts (V).

2. **Current (Water Flow)**

   Current is analogous to the flow of water through a pipe. Just as the water flow represents the amount of water passing through the pipe over a period of time, the electric current represents the amount of electric charge moving through a conductor or circuit over a period of time. Current is typically measured in amperes (A) or amps.

3. Resistance (Pipe Size)

Resistance in an electrical circuit can be compared to the size of the pipe in the water analogy. A pipe with a smaller diameter creates more resistance to water flow, making it more difficult for water to pass through. Similarly, resistance in a circuit represents the opposition or hindrance to the flow of electric current. The resistance depends on factors like the material of the conductor, its length, and its cross-sectional area. Resistance is typically measured in ohms (Ω).

3.3 Understanding of Ohm's Law

To design, install, and maintain an off-grid solar system, it's essential to have a clear understanding of the relationship between voltage, current, resistance, and power. These fundamental electrical concepts are interconnected, and their relationship can be expressed through several key formulas.

I. Ohm's Law:

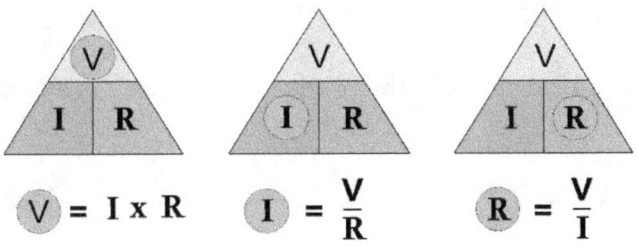

Fig 3.2

Ohm's Law describes the relationship between voltage (V), current (I), and resistance (R) in an electrical circuit:

$$V = I \times R$$

Where:

- V is the voltage (in volts)
- I is the current (in amperes)
- R is the resistance (in ohms)

Ohm's Law can be rearranged to solve for current or resistance:

$$I = V / R$$

$$R = V / I$$

The water analogy can also be used to understand Ohm's Law, the water flow (current) through a pipe is directly proportional to the water pressure (voltage) and inversely proportional to the size of the pipe (resistance). The greater the water pressure and the larger the pipe, the more water flows through it. Similarly, a higher voltage and lower resistance in a circuit result in a higher electric current flowing through it.

II. Power (P):

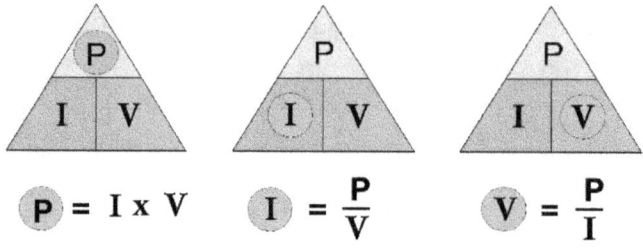

Fig 3.3

The power in an electrical circuit is the product of voltage (V) and current (I):

$$P = V \times I$$

Where:

- P is the power (in watts)
- V is the voltage (in volts)
- I is the current (in amperes)

III. Combining Ohm's Law and Power:

By combining Ohm's Law and the power equation, we can express power in terms of resistance and current, or voltage and resistance:

$$P = I^2 \times R$$

$$P = V^2 / R$$

Relationship between Power, Energy, and Time: Watt-hours and Kilowatt-hours

To design and size an off-grid solar system effectively, it is crucial to understand the relationship between power, energy, and time. This section will explore how watt-hours and kilowatt-hours are used to represent energy consumption and production in a solar system.

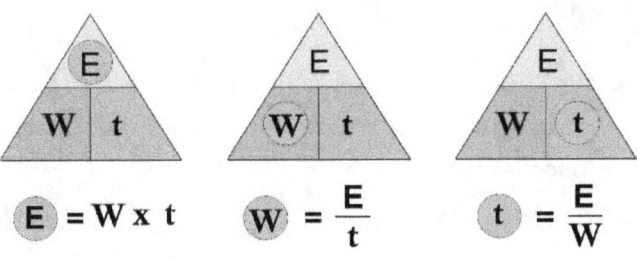

Fig 3.4

1. **Energy (E):** Energy, measured in watt-hours (Wh) or kilowatt-hours (kWh), represents the amount of electrical energy consumed or generated over time. It is the product of power (in watts or kilowatts) and time (in hours).

2. **Watt-hour (Wh):** A watt-hour is the amount of energy consumed or generated by a device with a power rating of one watt operating for one hour. For example, a 60-watt light bulb operating for 3 hours would consume 180 watt-hours of energy (60 W x 3 h = 180 Wh).

3. **Kilowatt-hour (kWh):** A kilowatt-hour is equal to 1,000 watt-hours. It is the standard unit of measurement for energy consumption and generation in both grid-tied and off-grid systems. For example, if a 1,000-watt appliance operates for 2 hours, it would consume 2 kilowatt-hours of energy (1 kW x 2 h = 2 kWh).

To calculate the energy consumption of an appliance or device in watt-hours or kilowatt-hours, use the following formula:

$$E \text{ (Wh)} = P \text{ (W)} \times t \text{ (h)}$$

$$E \text{ (kWh)} = P \text{ (kW)} \times t \text{ (h)}$$

Where:

*   E is the energy consumption or generation in watt-hours or kilowatt-hours

- P is the power rating of the appliance or device in watts or kilowatts
- t is the time during which the appliance or device operates, in hours

Understanding the relationship between power, energy, and time is essential for determining the energy requirements of your off-grid solar system. By calculating the energy consumption of each appliance and device in your system, you can estimate the total daily energy needs and appropriately size your solar array and battery bank.

3.4. Series and Parallel Connection

Series and parallel connections are two ways to arrange electrical components, such as resistors, capacitors, or batteries, in a circuit. Each type of connection has its distinct characteristics and applications.

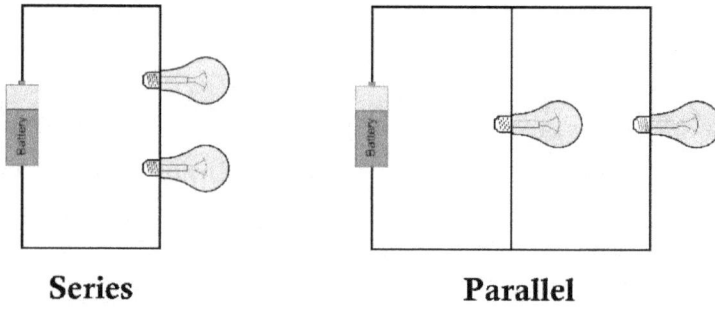

Series   **Parallel**

Fig 3.5

Series Connection:

In a series connection, components are connected end-to-end, forming a single path for the current to flow. The current remains the same throughout the entire path. If one component fails or is disconnected, the current will stop flowing through the entire circuit, much like a break in a chain.

Characteristics of Series Connection:

The total resistance (R_total) in the circuit is the sum of the individual resistances:

$$R\_total = R1 + R2 + R3 + ...$$

The total voltage (V_total) across the circuit is the sum of the individual voltage drops across each component:

$$V\_total = V1 + V2 + V3 + ...$$

The current (I) flowing through each component is the same.

Series connections are often used when it is necessary for the current to pass through each component in a specific sequence or when a voltage drop across each component is desired.

Parallel Connection:

In a parallel connection, components are connected side-by-side, forming multiple paths for the current to flow. The voltage across each path remains the same. If one component fails or is disconnected, the current will continue to flow through the remaining paths, much like the flow of water through multiple channels.

Characteristics of Parallel Connection:

The reciprocal of the total resistance ($1/R_{total}$) in the circuit is the sum of the reciprocals of the individual resistances:

$$1/R\_total = 1/R1 + 1/R2 + 1/R3 + ...$$

The total current (I_{total}) flowing through the circuit is the sum of the individual currents flowing through each component:

$$I\_total = I1 + I2 + I3 + ...$$

The voltage (V) across each component is the same.

Parallel connections are commonly used when components need to operate independently, and a constant voltage is required across each component.

3.5 Difference Between AC and DC

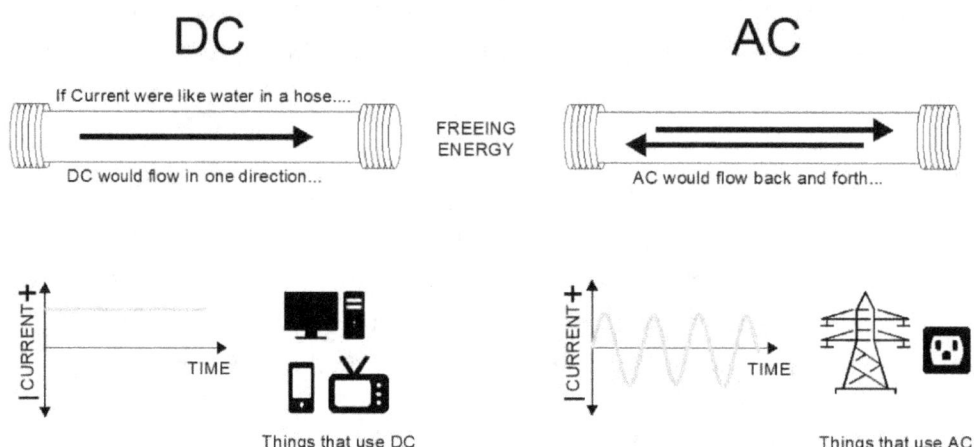

Fig 3.6

In a solar power system, you will likely have two types of electrical signals: DC (Direct Current) and AC (Alternating Current). Knowing the differences between these signals will help you figure out which devices to use, where to use them, and why they are necessary. Let's break it down:

DC (Direct Current):

DC is an electrical signal that always flows in one direction throughout the circuit. In a solar power system, solar panels generate DC electricity when they convert sunlight into energy. Batteries also store and supply electricity as DC.

AC (Alternating Current):

AC is an electrical signal that changes direction back and forth periodically. This is the type of electricity we use in our homes to power most of our appliances, like refrigerators, televisions, and computers.

You might be thinking, if AC signal keeps changing, how do we measure its value? Well, when we measure AC with tools like digital multimeters, they show us a special value known as the Root Mean Square (RMS). So, with a typical multimeter, you're usually seeing the RMS value, unless it says otherwise.

RMS, or Root Mean Square (RMS):

RMS, is a mathematical way to represent the "effective" or "average" value of an AC (alternating current) voltage or current. Since AC values oscillate between positive and negative, simply taking an average would result in a value of zero. Hence, the RMS value is used to get a meaningful average.

Here's a simplified explanation:

- Square all the instantaneous values of the AC waveform over one full cycle. Squaring makes all the values positive.
- Find the mean (average) of these squared values.
- Take the square root of that mean.

The RMS value gives a useful measure because it relates directly to the energy or power associated with the AC waveform. For many purposes, the RMS value of an AC waveform gives an equivalent DC (direct current) value. For example, an AC voltage with an RMS value of 120 volts will deliver the same amount of energy to a resistor as a DC voltage of 120 volts.

Why Do We Need Both Signals in a Solar Power System?

Solar panels produce DC electricity, but most of our household appliances run on AC electricity. To make the solar energy usable for these appliances, we need to convert the DC electricity from the solar panels into AC electricity. That is where a device called an inverter comes in. The inverter takes the DC electricity from the solar panels or batteries and turns it into AC electricity for use in your home.

3.6 Simple Math for Electrical Calculations

In designing and sizing an off-grid solar system, it's essential to be able to perform basic electrical calculations to determine system requirements and components. This section will introduce some simple mathematical formulas and examples that can be used in solar system design.

1. **Calculating Power (P):** Power is the product of voltage and current ($P = V \times I$). For example, if an appliance operates at 120 volts and draws 5 amps of current, its power consumption is 600 watts ($120 \text{ V} \times 5 \text{ A} = 600 \text{ W}$).

2. **Calculating Energy Consumption (E):** Energy consumption is the product of power and time ($E = P \times t$). For example, if a 600-watt appliance operates

for 3 hours, it would consume 1,800 watt-hours of energy (600 W x 3 h = 1,800 Wh).

3. **Calculating Total Daily Energy Consumption:** To estimate the total daily energy consumption of your off-grid solar system, calculate the energy consumption for each appliance and add them together. For example:

    - Appliance A: 200 W, used for 4 hours daily = 800 Wh
    - Appliance B: 100 W, used for 6 hours daily = 600 Wh
    - Appliance C: 50 W, used for 8 hours daily = 400 Wh

    Total daily energy consumption = 800 Wh + 600 Wh + 400 Wh = 1,800 Wh or 1.8 kWh

4. **Calculating Current from Power and Voltage:** Using Ohm's Law (I = V / R) and the power equation (P = V x I), we can calculate the current (I) in a circuit when given the power (P) and voltage (V):

$$I = P / V$$

    For example, if an appliance consumes 600 watts of power at a voltage of 120 volts, it draws a current of 5 amps (600 W / 120 V = 5 A).

5. **Calculating Resistance from Voltage and Current:** Using Ohm's Law (R = V / I), we can calculate the resistance (R) in a circuit when given the voltage (V) and current (I):

$$R = V / I$$

    For example, if the voltage across a resistor is 12 volts, and the current flowing through it is 2 amps, the resistance is 6 ohms (12 V / 2 A = 6 Ω).

By applying these simple mathematical formulas, you can perform essential electrical calculations to design and size your off-grid solar system.

3.7 Energy Cost

Every state or country has different rates for electricity. To determine how much, you owe your electricity provider for usage, it's essential to be aware of your local electricity rate.

If we take the U.S. national average rate of $0.12 per kilowatt-hour as an example and you run a LED light bulb with a power rating of 20 Watts for 12 hours daily over a span of 30 days, here's how to compute the cost:

1. Determine Daily Energy Consumption:

Energy (in watt-hours) = Power (in watts) × Time (in hours)

$$= 20W \times 12h = 240Wh$$

This means the light utilizes 240 watt-hours each day.

2. Estimate Monthly Energy Consumption

Monthly Energy (in watt-hours) = Daily Energy (in watt-hours) × Number of days

$$= 2400Wh \times 30days = 7200Wh$$

This is equivalent to 7200 watt-hours or 7.2 kilowatt-hours (as 1 kilowatt-hour = 1000 watt-hours).

3. Calculate Monthly Cost

You can easily determine how much; you owe your electricity provider by using this simple math

Monthly Electricity Bill (in $) = Monthly Energy (in kilowatt-hours) x Electricity Rate (in $/kWh)

$$= 7.2kWh \times \$0.12/kWh = \$0.864$$

So, using a 20-Watt light for 12 hours each day over 30 days will lead to a cost of $0.864.

3.8 Digital Multimeter

A Digital Multimeter (DMM) is a multi-functional instrument that consolidates various metering functions into a single unit. It primarily measures electrical values such as voltage (volts), current (amps), and resistance (ohms). It is a standard diagnostic tool for technicians in the electrical/electronic industries.

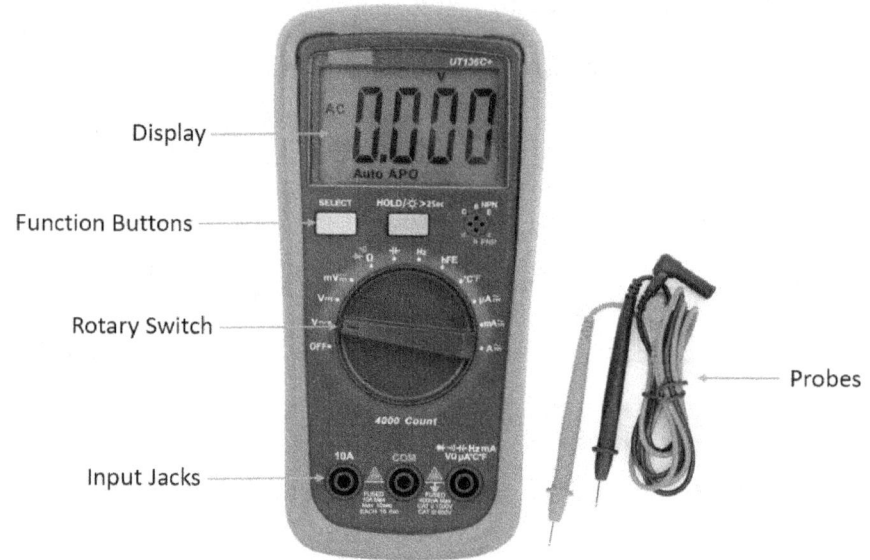

Fig 3.7

Important Parts of a Multimeter :

A digital multimeter is a handy tool that combine the testing capabilities of single-task meters - the voltmeter (for measuring volts), ammeter (amps) and ohmmeter (ohms). In addition, many multimeters have extra features for different tasks.

When you look at a digital multimeter, you will usually see these parts:

- **Display:** This is where you read the measurements.

- **Function Buttons:** These let you choose different functions. Each multimeter will have different options.

- **Rotary Switch:** For selecting primary measurement values (volts, amps, ohms) and their range

- **Input Jacks:** This is where you plug in the tester cables.

- **Probes / Leads:** These are the cables you insert into the input jacks. One is usually black (negative) and the other red (positive), and they're used to connect the multimeter to whatever you're testing.

How to Use a Multimeter?

1. Measuring Voltage

It measures the potential difference between two points in a circuit. Capable of measuring both AC (Alternating Current) and DC (Direct Current) voltages.

Step-1: Set Up the Leads

- Insert the black lead into the slot labeled "COM" (Common).
- Fit the red lead into the "V" slot, indicating Voltage.

Step-2: Select AC or DC Voltage

- For DC voltage (typically in batteries or solar panel), shift the dial to the V sign followed by a straight line ($=$)
- For AC voltage (commonly found in household electrical items like lamps or blenders), turn the dial to the V symbol accompanied by a wave ($\sim$).

Step-3: Choose the Appropriate Voltage Range

- Always select a range higher than the anticipated voltage to ensure accuracy.
- For instance, if you're assessing a 12V battery and your multimeter has settings for 2V and 20V, go with 20V.
- If uncertain about the voltage, it's safest to use the highest setting on your multimeter.

Step-4: Read the Display

Once everything is set up, the multimeter will display the measured Voltage

2. Measuring Current

When measuring current using a multimeter, it's crucial to differentiate between milliamperes (mA) and Amperes (A) as they represent different scales of current. Here's a revised guide for measuring both:

Step-1: Setting Up the Leads

- Plug the black test lead into the jack labeled "COM" (Common).
- For milliampere (mA) measurements, insert the red test lead into the "mA" jack.
- For measurements up to 10 Amperes, plug the red test lead into the "10A" jack.

Step-2: Select AC or DC Current

- For DC current, shift the dial to the A or mA sign followed by a straight line (=).
- For AC current, turn the dial to the A or mA symbol accompanied by a wave (~).

Step-3: Integrate Multimeter into the Circuit

- To measure the current, you'll need to make the multimeter a part of the circuit. This involves breaking open the circuit at the point you intend to measure and connecting the multimeter in series.

Step-4: Determine the Appropriate Current Range

- If you're expecting a low current, use the mA setting on the multimeter.
- For currents close to or above 1 Ampere, use the 10A setting. Always start with a higher range if unsure, then dial down to get an accurate reading without risking overloading the multimeter.

Step-5: Read the Display

- Once everything is correctly set, the multimeter will display the current value flowing through that part of the circuit

Note: Using a digital multimeter to measure current is possible, but an clampmeter is often more suitable. When measuring with a multimeter, current flows directly through it, which can risk damage or blow its internal fuse if there's no applied load. Many meters have a limit of around 10Amps, which isn't very high. Instead of directly measuring, you can also determine the current by applying the formulas we'll explore in the next section.

3. Measuring Resistance

Step-1: Setting Up the Leads

- Plug the black test lead into the terminal labeled "COM" (Common).
- Plug the red test lead into the terminal usually marked with the ohm (Ω) symbol.

Step-2: Select Resistance Mode

Turn the dial of the multimeter to the resistance (Ω) setting. This symbol is usually an uppercase Greek letter omega (Ω).

Step-3: Choose the Appropriate Range

- Most modern multimeters are auto-ranging, meaning they automatically find the correct resistance range for what you're testing. However, if yours isn't: Set it to the lowest resistance range initially.
- If the multimeter reads '1' or displays 'OL' (overload), it means the resistance is too high for that setting. In this case, turn the dial to a higher resistance range until you get a reading.

Step-4: Test the Component

- Touch the probes to the component terminals. The polarity doesn't matter for resistance measurements.
- Wait a moment for the reading to stabilize, and then note the displayed resistance value.

I personally own the UNI-T UT 136C+ multimeter which is an excellent tool for both DIY projects and solar setups. Its compact size makes it easy to handle and carry, ideal for those who are always on the move. For solar applications, it provides fairly accurate AC/DC voltage readings. Its auto-ranging capability means no time is wasted on manual adjustments, ensuring quick and precise readings. Moreover, the clear backlit display ensures easy visibility, whether you're working indoors or out in the field.

3.9 Ammeter or Clamp Meter

A clamp meter, often referred to as a "clamp-on ammeter", is a specialized electrical tool designed to measure current without needing to make direct contact with the conductor. Unlike a traditional multimeter which requires the circuit to be broken to measure current, a clamp meter simply clamps around a conductor, allowing for quick and safe measurements. This is possible because the clamp utilizes the magnetic field created by the current as it flows through a conductor.

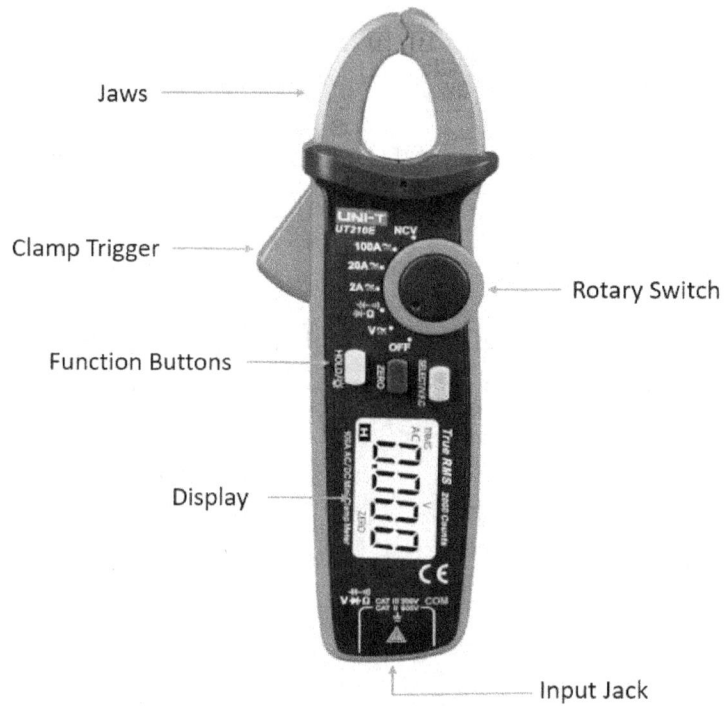

Fig-3.8

A clamp meter, offers a distinct advantage over traditional multimeters when it comes to measuring current. Its unique design incorporates a clamp, allowing you to measure the RMS (root mean square) value of an electrical current without breaking the circuit. To use it, you simply open the clamp, encircle a single conductor, and read the current flow. It's essential to remember that if you try to measure a cable containing both positive and negative wires, the readings will neutralize each other, giving a zero or inaccurate result.

Budget-friendly clamp meters typically measure only AC current. If you're looking to measure DC current, ensure the one you're considering is capable of doing so.

Additionally, be mindful of its ampere rating; for most systems, a capacity of at least 100Amps is recommended.

I personally own an UNI-T UT210E clamp meter which is budget friendly and a solid choice for both DIY enthusiast and professionals. Its small and portable design makes it easy to use in tight spaces and carry around, especially in off-grid locations where you might not have the luxury of a full workshop. This clamp meter can measure both AC and DC current.

How to use a Clamp meter?

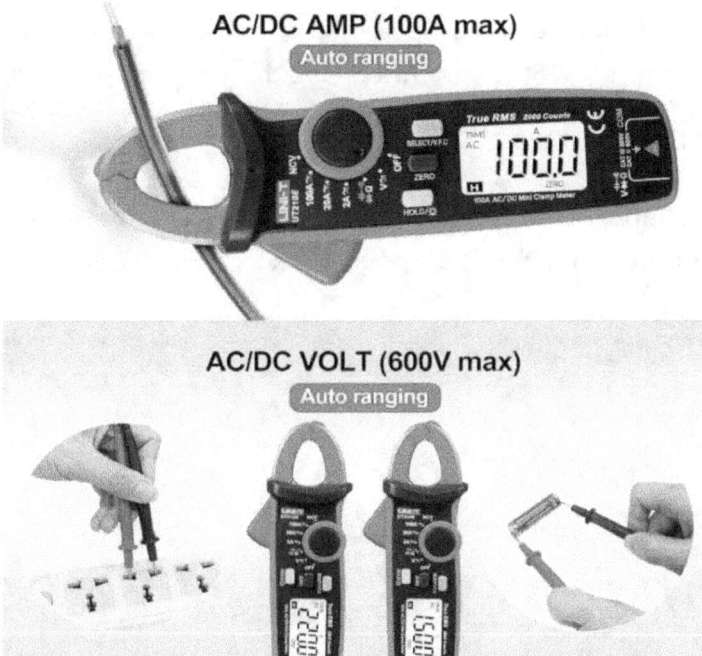

Fig-3.9

Using a clamp meter is fairly straightforward. Just follow these simple steps given below:

Step-1: Turn On the Meter

- Start by turning on the clamp meter using the power button.

Step-2: Select the Measurement Type

- Choose what you want to measure, like AC or DC current. Rotate the selection dial to the desired setting.

Step-3: Open the Clamp

- Press the trigger or lever to open the jaws of the clamp meter.

Step-4: Clamp Around a Conductor

- Place only one wire inside the jaws. Ensure you don't clamp around both positive and negative wires together, as their readings will cancel each other out.

Step-5: Read the Measurement

- Once the conductor is inside the jaws, the meter will give a reading on its display. This shows the current flowing through the conductor.

Step-6: Zero Out

- Some clamp meters have a 'zero' or 'null' button. If your reading seems off, press this button with the clamp in open air (not around a conductor) to reset it to zero.

3.10 Energy Meter

Energy meters tell us how much electricity we use, so we know what to pay and can try to use less.

In the off-grid solar system, energy meter is installed after the solar panels and before the battery or load. This placement allows the meter to measure the energy generated or consumed.

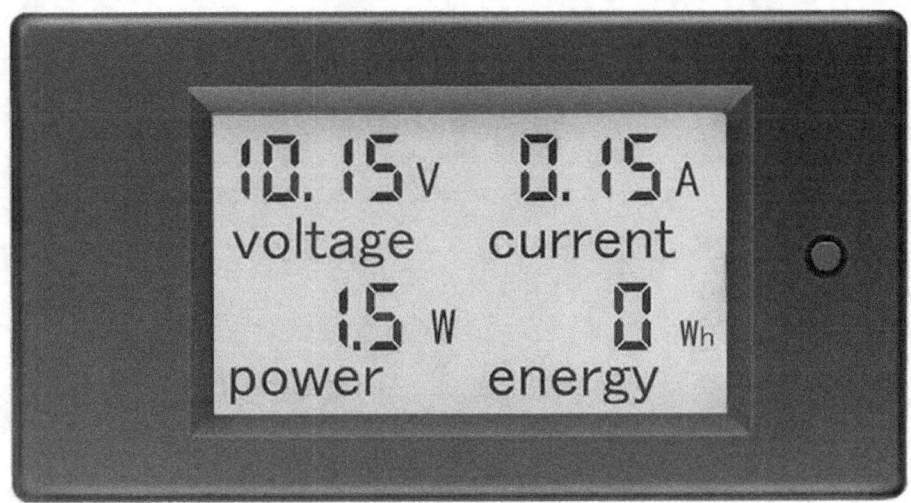

Fig-3.10

When you glance at the DC Energy meter display, you might see the following readings:

Voltage (V):

This indicates the current voltage level of the DC electricity. For a typical residential PV system, this might read somewhere between 12V to 48V, depending on the setup.

Current (A):

This shows the electrical current, in amperes, being generated. A reading might be "8.3A", denoting the solar panels are currently outputting 8.3 amperes of current.

Power (W):

This shows the current power output from the solar panels. For instance, on a sunny afternoon, it might read "250W" indicating the panels are currently producing 250 watts of power.

Accumulated Energy (Wh or kWh):

This measures the total energy produced over a given period. For example, it might display "5.2 kWh", signifying that the solar panels have generated 5.2 kilowatt-hours of energy since the last reset or from the start of the day.

Chapter 4
Solar Batteries

Off-grid solar systems generate electricity during the day, but this power supply is inconsistent due to variations in sunlight caused by weather conditions or the day-night cycle. Batteries are essential components in off-grid solar systems because they address the intermittent nature of solar power generation, ensuring a continuous and reliable supply of electricity. By storing excess solar energy produced during the day, batteries can provide power during periods when solar production is low, such as during the night or on cloudy days.

The primary function of batteries in off-grid solar systems is to store electrical energy and deliver it when needed. This energy storage capability is essential for maintaining a stable power supply and avoiding blackouts or power disruptions. Batteries also help in optimizing the use of solar energy by allowing the system to store surplus energy during peak production periods and release it during peak consumption periods or when solar energy production is low.

4.1 Types of Lead Acid Battery

Lead-acid batteries are classified into two main types: Flooded Lead-Acid (FLA) and Sealed Lead-Acid (SLA). Each type has its unique construction characteristics, advantages, and disadvantages.

Fig 4.1

Source: Exide Battery

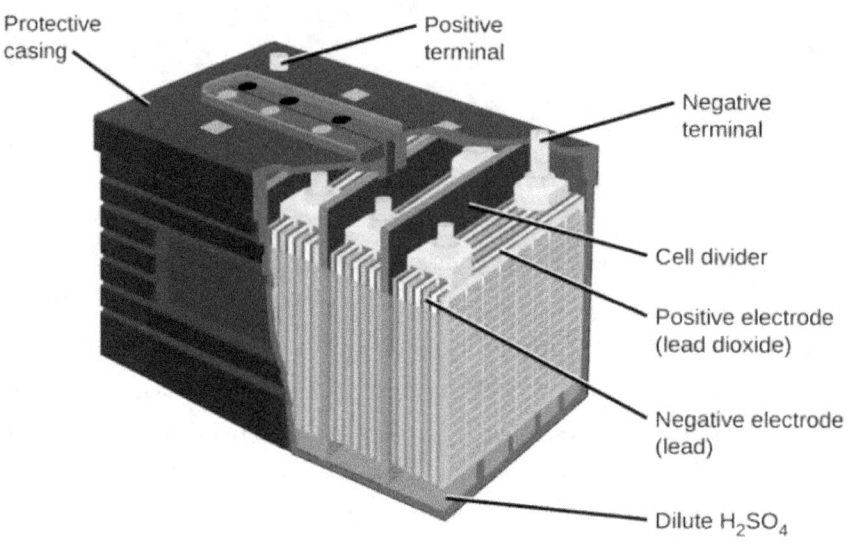

Fig 4.2

Source: Open Stax

1. Flooded Lead-Acid (FLA) Batteries:

Flooded lead-acid batteries, also known as wet cell or vented batteries have been the most common type of lead-acid battery for decades. In these batteries, the lead plates are submerged in a liquid electrolyte solution of sulphuric acid and water.

These batteries contain liquid electrolyte solution, consisting of sulfuric acid and water, in which the lead plates are submerged. The construction of FLA batteries includes positive and negative plates made of lead dioxide (PbO2) and sponge lead (Pb), respectively, which are suspended in the electrolyte solution. To prevent short circuits, porous separators made from materials like rubber, PVC, or microporous polyethylene are placed between the positive and negative plates. The battery cells, plates, and electrolyte are housed in a durable container, typically made of hard plastic. Each cell has a vent cap that allows gases to escape during the charging process and prevents the ingress of contaminants.

The exploded diagram of flooded lead acid battery is shown below.

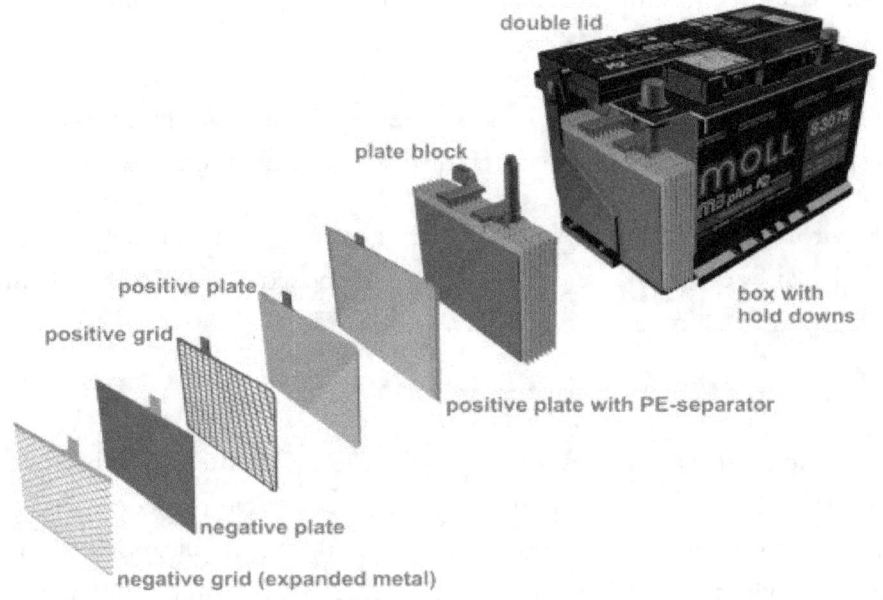

Fig 4.3

Source: Elsevier B.V

Advantages of lead-acid batteries:

- Affordability: FLA batteries are generally more cost-effective compared to other battery types, making them a popular choice for off-grid solar systems with budget constraints.

- Availability: Flooded Lead-acid batteries are widely available in various sizes and capacities, making it easy to find a suitable option for a specific off-grid solar system

- Proven performance: With over a century of use in various applications, lead-acid batteries have a well-established track record for reliability and performance.

Disadvantages of lead-acid batteries

- Maintenance: Flooded lead-acid batteries require regular maintenance, such as topping up with distilled water and checking electrolyte levels. This can be time-consuming and may not be suitable for remote or unattended installations.

- Lower energy density: Lead-acid batteries have a lower energy density compared to lithium-ion batteries, which means they require more space to store the same amount of energy.

- Shorter lifespan: They have a shorter lifespan than lithium-ion batteries, especially when subjected to deep discharges or high temperatures.

- Environmental concerns: Due to the presence of lead and sulphuric acid, lead-acid batteries must be handled and disposed of carefully to minimize environmental impact.

2. Sealed Lead-Acid (SLA) Batteries:

Sealed Lead-Acid (SLA) batteries are a type of lead-acid battery that has been sealed and designed to be maintenance-free. They are an attractive choice for off-grid solar systems, uninterruptible power supplies (UPS), and other applications where regular maintenance may not be feasible. Sealed Lead-Acid (SLA) batteries are also known as VRLA or valve-regulated lead-acid. They have a pressure-sensitive valve that automatically controls the emission of gases, but in normal operation conditions, they are closed. They are opened automatically to release gases in case there is high pressure inside the battery if there is something wrong with the battery, like a short circuit.

SLA batteries come in two main subtypes: Absorbent Glass Mat (AGM) and Gel.

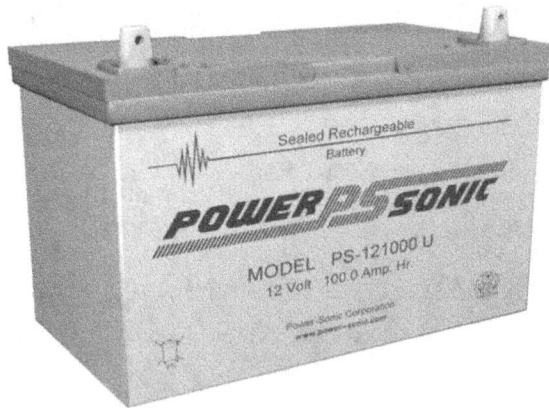

Fig 4.4

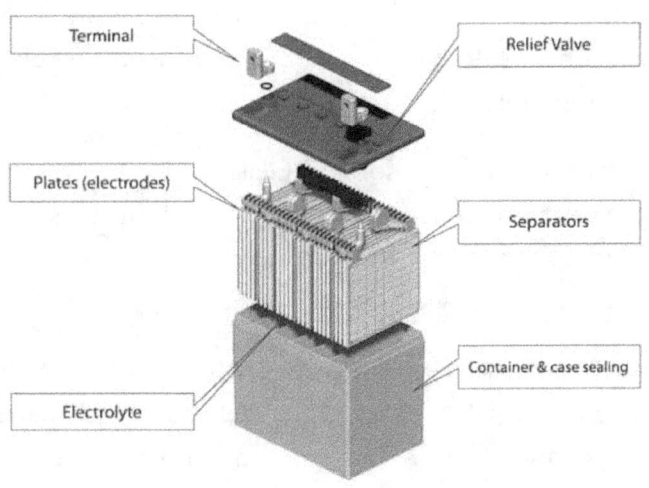

Fig 4.5

Source: power sonic

- **Absorbent Glass Mat (AGM) batteries:**

AGM batteries are constructed with positive and negative plates made of lead dioxide ($PbO2$) and sponge lead (Pb), similar to Flooded Lead-Acid (FLA) batteries. However, instead of a liquid electrolyte, AGM batteries use a highly porous fiberglass separator that absorbs the electrolyte, making the battery spill-proof and more resistant to vibration. This fiberglass mat helps maintain close contact between the electrolyte and the plates, improving the battery's performance.

- **Gel Batteries:**

Gel batteries also feature alternating positive and negative plates made of lead dioxide and sponge lead. The key difference between Gel batteries and other lead-acid batteries is the use of a gel-like electrolyte. The gel electrolyte is created by adding silica to the sulphuric acid solution, forming a thick, non-spillable gel that prevents leakage and allows the battery to function in various orientations.

Advantages of Sealed Lead-Acid (SLA):

Sealed Lead-Acid (SLA) batteries offer several advantages over Flooded Lead-Acid (FLA) batteries, making them a popular choice for various applications, including off-grid solar systems. Some of the key advantages of SLA batteries over FLA batteries are:

35

Maintenance-free: SLA batteries are sealed and do not require regular maintenance tasks such as topping up with distilled water or equalization charging. This feature makes them a more convenient option for applications where regular maintenance may be challenging or not feasible.

Spill-proof and leak-proof: Both Absorbent Glass Mat (AGM) and Gel SLA batteries have immobilized electrolytes, which eliminates the risk of spills or leaks. This feature makes SLA batteries safer to use in various orientations and in situations where liquid electrolyte spills may cause damage or pose a hazard.

No off-gassing: During the charging process, FLA batteries produce hydrogen and oxygen gases, which are vented through the battery caps. SLA batteries, on the other hand, have a recombination process that prevents most of the gases from escaping, making them a more environmentally friendly option.

Better performance in extreme temperatures: SLA batteries, especially AGM batteries, usually perform better in extreme temperature conditions compared to FLA batteries, making them suitable for use in harsh environments.

Lower self-discharge rate: SLA batteries have a lower self-discharge rate compared to FLA batteries, allowing them to retain their charge for more extended periods when not in use. This characteristic makes them ideal for applications with irregular charging patterns or where long periods of inactivity are expected.

Despite these advantages, it is essential to note that SLA batteries are more expensive than Flooded Lead-Acid (FLA) batteries.

4.2 Types of Lithium Battery

Lithium-ion batteries have gained popularity in recent years due to their superior performance, higher energy density, and longer lifespan compared to traditional lead-acid batteries. They are increasingly being used in off-grid solar systems, electric vehicles, and consumer electronics.

Fig 4.6

Source: Greensun Solar

Construction and working principle:

Lithium-ion batteries are composed of multiple cells that contain a positive electrode (cathode), a negative electrode (anode), an electrolyte, and a separator. The cathode is typically made from lithium metal oxide, while the anode is made from carbon, often in the form of graphite. The electrolyte is a lithium salt solution in an organic solvent, and the separator is a thin porous membrane that keeps the electrodes from touching each other while allowing the passage of lithium ions.

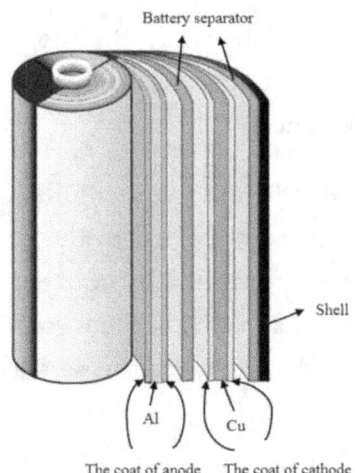

Fig 4.7

During the discharge process, lithium ions move from the anode to the cathode through the electrolyte, generating an electric current. When the battery is recharged, lithium ions move back from the cathode to the anode.

There are several chemistries of lithium-ion batteries, with each offering different performance characteristics. Some common lithium-ion chemistries include:

- **Lithium Iron Phosphate (LiFePO4):** LiFePO4 batteries are known for their long cycle life, thermal stability, and safety. They are often preferred for solar energy storage applications due to their durability and lower risk of thermal runaway.

- **Lithium Cobalt Oxide (LiCoO2):** LiCoO2 batteries have a high energy density, making them popular in portable electronics such as smartphones and laptops. However, they have lower thermal stability and are more prone to thermal runaway, making them less suitable for large-scale energy storage applications.

- **Lithium Manganese Oxide (LiMn2O4):** LiMn2O4 batteries offer a good balance between energy density, power output, and safety. They are often used in power tools and electric vehicles.

- **Lithium Nickel Manganese Cobalt Oxide (LiNiMnCoO2 or NMC):** NMC batteries provide high energy density and a long cycle life, making them suitable for electric vehicles and some stationary energy storage applications.

4.3 Lithium Iron Phosphate (LiFePO4 or LFP) Batteries

LFP are a type of lithium-ion battery that has gained popularity due to their safety, long cycle life, and excellent thermal stability. LFP batteries are widely used in electric vehicles, renewable energy systems, and various other applications. Here are some details and varieties of LFP batteries:

LFP batteries have a similar construction to other lithium-ion batteries, consisting of a positive electrode (cathode), a negative electrode (anode), an electrolyte, and a separator. The cathode is made from lithium iron phosphate (LiFePO4), while the anode is typically made from carbon, often in the form of graphite. The electrolyte is a lithium salt solution in an organic solvent, and the separator is a thin porous membrane that keeps the electrodes from touching each other while allowing the passage of lithium ions.

During the discharge process, lithium ions move from the anode to the cathode through the electrolyte, generating an electric current. When the battery is recharged, lithium ions move back from the cathode to the anode.

Varieties of LFP Batteries:

Lithium Iron Phosphate (LFP) batteries are available in different form factors and types, catering to various applications and requirements. The main types of LFP batteries are based on their cell design, which includes:

Prismatic LFP cells:

Fig 4.8

Prismatic LFP cells are rectangular-shaped cells that are popular in electric vehicles and large-scale energy storage systems due to their high energy density and modular design. Prismatic cells are encased in a hard aluminum or steel casing that provides structural support and helps dissipate heat during operation. These cells can be easily assembled into battery packs by stacking and connecting them in series or parallel configurations to achieve the desired voltage and capacity.

Cylindrical LFP cells:

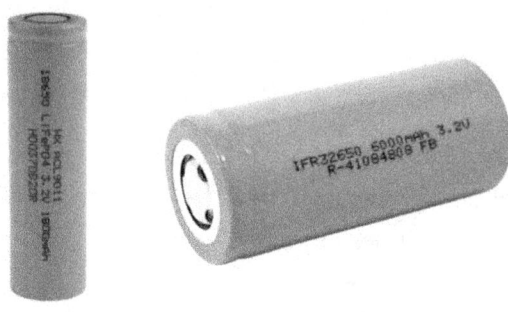

Fig 4.9

Cylindrical LFP cells have a cylindrical shape and are commonly used in portable electronics, power tools, and smaller-scale energy storage systems. They come in various sizes, such as 18650, and 32650, with the numbers indicating the cell's diameter and length in millimetres. Cylindrical cells have a steel or aluminium casing, providing structural support and protection.

Pouch LFP cells:

Fig 4.10

Pouch LFP cells feature a flexible and lightweight packaging, typically made from laminated aluminum and polymer films. This design allows pouch cells to be lightweight and occupy less space, making them ideal for portable electronics and electric vehicles, where weight and space are crucial factors. Pouch cells can be assembled into battery packs by stacking and connecting them in series or parallel configurations, and they often require additional support structures or enclosures to maintain their shape and provide protection.

Custom LFP battery packs:

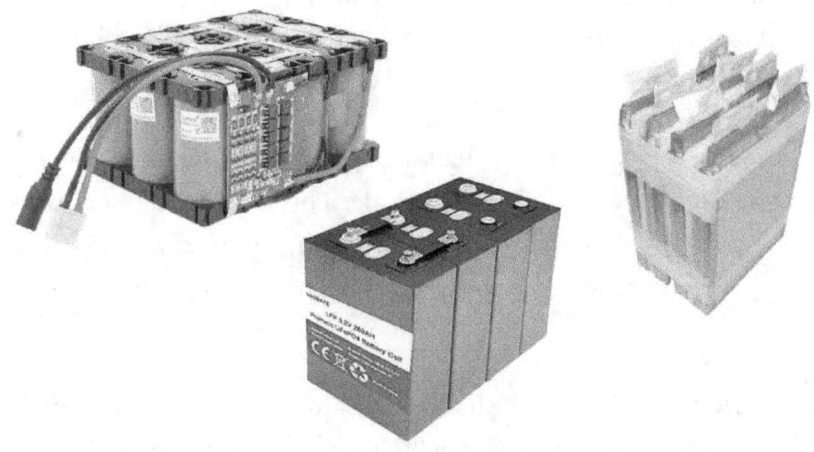

Fig 4.11

Custom LFP battery packs are assembled using prismatic, cylindrical, or pouch cells to meet specific voltage, capacity, and form factor requirements for various applications. Custom battery packs may incorporate built-in battery management systems (BMS), thermal management systems, and protection circuits, ensuring the safe and efficient operation of the battery.

Advantages of LFP Batteries:

- **Safety:** LFP batteries have a more stable chemistry compared to other lithium-ion batteries, reducing the risk of thermal runaway and making them a safer choice.

- **Long cycle life:** LFP batteries typically have a longer cycle life than other lithium-ion chemistries, making them suitable for applications where the battery undergoes frequent charge and discharge cycles.

- **High thermal stability:** LFP batteries can operate in a wide temperature range without significant degradation, making them suitable for use in various environments.

- **Environmentally friendly:** LFP batteries do not contain toxic heavy metals like cobalt or nickel, making them a more environmentally friendly choice.

Disadvantages of Lithium Batteries:

- **Cost:** Lithium-ion batteries tend to be more expensive than other types of batteries such as lead-acid. This is due to the high cost of lithium and other materials used in their construction.

- **Safety Concerns:** If improperly handled or charged, lithium-ion batteries can overheat and potentially catch fire or explode. This is due to a phenomenon known as thermal runaway. However, the risk is significantly lower with Lithium Iron Phosphate (LFP) batteries, which are considered among the safest lithium-ion chemistries.

- **Requires Protection Circuit:** Lithium-ion batteries require a protection circuit to prevent overcharging or discharging, which can damage the battery and pose safety risks. This adds to the complexity and cost of the battery system.

- **Sensitivity to High Temperatures:** Lithium-ion batteries can degrade faster if exposed to high temperatures. This can be a limitation in applications or environments where high temperatures can't be avoided.

4.4 Battery State of Charge (SoC)

The state of charge (SoC) is defined as the amount of energy in a battery, expressed as a percentage of the energy stored in a fully charged battery. Discharging a battery results in a decrease in state of charge, while charging results in an increase in state of charge. If a battery's SoC is 100%, it's fully charged, while an SoC of 0% means it's fully discharged.

SoC is typically calculated based on the current output of a battery and its known capacity.

For example, if a battery has a capacity of 100 ampere-hours (Ah) and it has supplied 20 Ah of current, its SoC would be (100-20)/100 * 100% = 80%.

Below are the complete voltage chart for Flooded Lead Acid (FLA), Sealed Lead Acid (SLA), Lithium Iron Phosphate (LFP) batteries. Please note, these are approximate values and can vary based on several factors such as the specific battery design, age, temperature, and usage history. Always refer to the manufacturer's specifications for accurate information.

12V Sealed Lead Acid Battery Voltage Chart	
Voltage	Capacity
12.9	100%
12.8	90%
12.6	80%
12.5	70%
12.4	60%
12.2	50%
12.1	40%
11.9	30%
11.8	20%
11.7	10%
11.6	0%

12V Floaded Lead Acid Battery Voltage Chart	
Voltage	Capacity
12.7	100%
12.5	90%
12.4	80%
12.3	70%
12.2	60%
12	50%
11.9	40%
11.8	30%
11.7	20%
11.6	10%
11.5	0%

Fig 4.12

Lithium Iron Phosphate (LiFePO4) Battery Voltage Chart			
1Cell (3.2V)	4 Cells (12V)	8 Cells (24V)	Capacity
3.65	14.6	29.2	100%
3.35	13.4	26.8	90%
3.32	13.3	26.6	80%
3.3	13.2	26.4	70%
3.27	13.1	26.1	60%
3.26	13	26.1	50%
3.25	13	26	40%
3.22	12.9	25.8	30%
3.2	12.8	25.6	20%
3.0	12.0	24.0	10%
2.6	10.4	20.8	0%

Fig 4.13

4.5 Battery Depth of Discharge (DoD)

DoD stands for Depth of Discharge and is defined as the percentage of capacity that has been withdrawn from a battery compared to the total fully charged capacity. By definition, the depth of discharge and state of charge of a battery add to 100 percent.

It is the opposite of State of Charge (SoC). If a battery's SoC tells us how much energy it has left, its DoD tells us how much it has already used.

Similar to SoC, DoD can be calculated based on the current output of a battery and its known capacity. For instance, if a battery has a capacity of 100 ampere-hours (Ah) and it has supplied 20 Ah of current, its DoD would be $(20/100) * 100\% = 20\%$.

DoD and Battery Types:

The acceptable DoD varies between different types of batteries:

Lead-acid batteries typically have a recommended DoD of around **50%**. Regularly discharging beyond this level can significantly reduce their lifespan.

Lithium-ion batteries can typically handle higher DoD levels than lead-acid batteries. Some lithium-ion batteries can be regularly discharged to a DoD of **80%** without significantly impacting their lifespan. Lithium Iron Phosphate (LiFePO4) batteries are often rated for 2000 to 5000 cycles at 80% DoD.

4.6 Battery Cycle Life

A single "cycle" for a battery involves discharging the battery from its full capacity down to a certain level (Depth of Discharge or DoD), and then recharging it back up to full capacity. The cycle life is the number of these charge-discharge cycles a battery can undergo before its performance degrades to a specified level.

The DoD vs Cycle life curve helps in understanding how the lifespan of a battery (measured in cycles) changes with different depths of discharge.

In general, a battery's cycle life decreases as the DoD increases. This means that if a battery is regularly discharged deeply (high DoD), it won't last as many cycles compared to when it's discharged less deeply (low DoD). This is because the deeper a battery is discharged in each cycle, the more wear and tear it experiences, leading to a shorter overall lifespan.

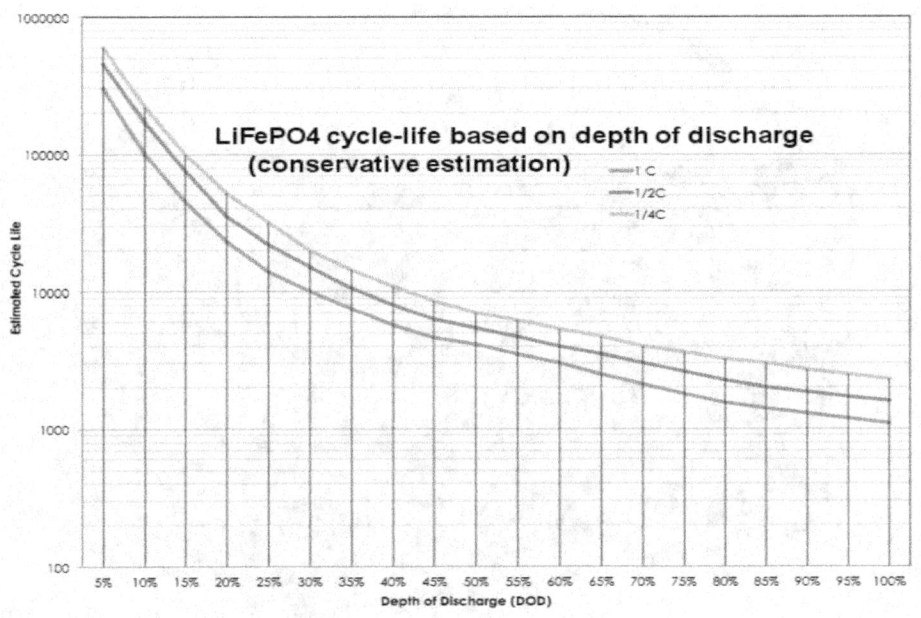

Fig 4.14

Source: gwl-power

4.7 Battery C Rating

The C rating is defined as the charge or discharge current divided by the battery's capacity. For example, if a battery with a capacity of 100 ampere-hours (Ah) is charged or discharged at a rate of 50 amperes, its C rate is $50/100 = 0.5C$.

C ratings are often used to specify both the maximum safe continuous discharge rate of a battery and the rate at which it can be safely charged.

The formula to calculate C rating is quite straightforward:

$$\textbf{C Rating = Current / Capacity}$$

Where:

- Current is the charge or discharge current, typically measured in amperes (A).
- Capacity is the battery's capacity, typically measured in ampere-hours (Ah).

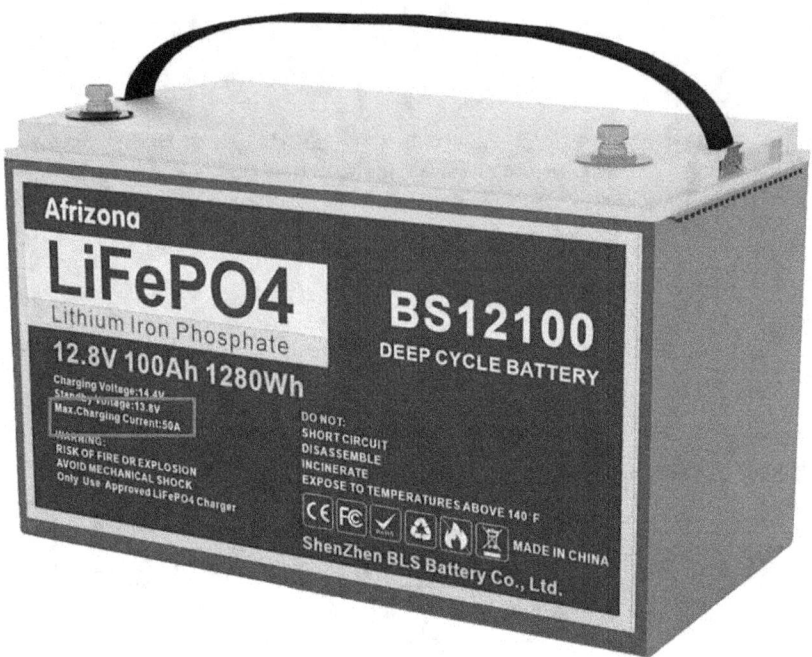

Fig 4.15

For example, in the above battery with a capacity of 100 Ah is being charged or discharged at a rate of 50 A, then its C rating would be:

C Rating = 50 A / 100 Ah = 0.5C

This means the battery is being charged or discharged at a rate equal to half its capacity per hour. If the current were 100 A instead, the C rating would be 1C, meaning the battery is being charged or discharged at a rate equal to its full capacity per hour.

C Rating and Battery Types:

Different types of batteries can handle different C rates:

- Lead-acid batteries: These batteries generally prefer lower C rates. A typical charge rate might be 0.1C to 0.2C, and a typical maximum discharge rate might be 1C to 2C.

- Lithium-ion batteries: These batteries can typically handle higher C rates than lead-acid batteries. Charge rates of 0.5C to 1C and discharge rates of 1C to 2C are common, although some types can handle higher rates.

C Rating and Battery Life:

The C rating can impact the lifespan of a battery. Charging or discharging a battery at a high C rate can cause it to heat up, which can accelerate capacity loss and reduce the battery's lifespan. Therefore, it's generally better for the battery's longevity to charge and discharge it at lower C rates.

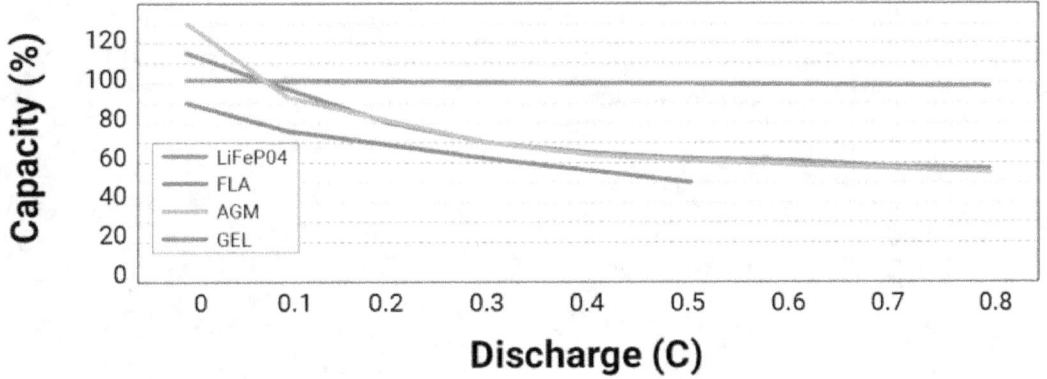

Capacity of lithium battery vs lead acid at various discharge currents

Fig 4.16

Source: Power Sonic

4.8 Battery Charging

A. Lead Acid Battery

Charging a lead-acid battery involves several stages to ensure that the battery is charged efficiently and safely. These stages are typically referred to as the bulk stage, the absorption stage, and the float stage. Some chargers also include an equalization stage.

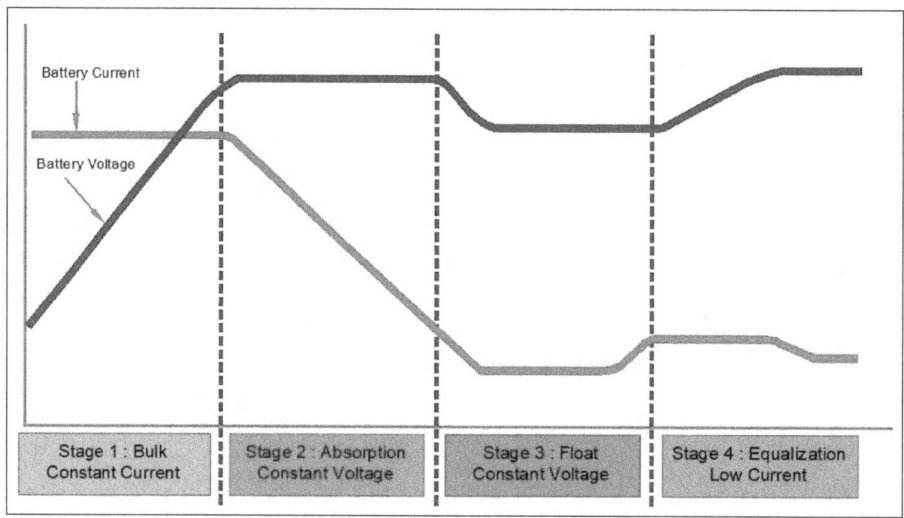

Fig 4.17

1. Bulk Stage:

The bulk stage is the first stage of charging a lead-acid battery. During this stage, the charger provides a constant current to the battery. The goal is to recharge the battery as quickly as possible without exceeding its maximum safe charging rate (usually specified by the manufacturer).

As the battery charges, its voltage gradually increases. Once the voltage reaches a set point, typically around 14.4 volts for a 12-volt battery, the charger moves on to the absorption stage. The bulk stage usually accounts for about 80% of the battery's recharge.

2. Absorption Stage:

During the absorption stage, the charger maintains a constant voltage, and the charging current gradually decreases. This stage is designed to fully charge the battery without causing overheating or excessive gassing.

The charger stays in the absorption stage until the charging current drops to a predetermined level, indicating that the battery is fully charged. This stage usually accounts for the remaining 20% of the battery's recharge.

3. Float Stage:

Once the battery is fully charged, the charger enters the float stage. During this stage, the charger reduces the voltage to a lower level, typically around 13.2 to 13.8 volts for a 12-volt battery. This allows the battery to stay fully charged while minimizing gassing and water loss.

4. Equalization Stage (if applicable):

Equalization is an occasional, controlled overcharge that helps to balance the individual cells in a battery, preventing weak cells from dragging down the performance of the entire battery. It also helps to remove sulphate crystals that can build up on the battery's plates over time. Not all chargers include an equalization stage, and it's not recommended for all types of lead-acid batteries.

B. Lithium Iron Phosphate (LiFePO4 or LFP) Battery

The charging process for Lithium Iron Phosphate (LiFePO4 or LFP) batteries, like other lithium-ion batteries, typically involves two main stages: the constant current stage (CC) and the constant voltage stage (CV). Some battery management systems may also include a balancing stage.

1. Constant Current (CC) Stage:

During the constant current stage, the charger supplies the battery with a constant current. The battery's voltage gradually increases during this stage. The constant current phase allows for rapid charging and typically continues until the battery voltage reaches its peak value (usually around 3.6V per cell, or 14.4V for a 4-cell battery in a 12V system).

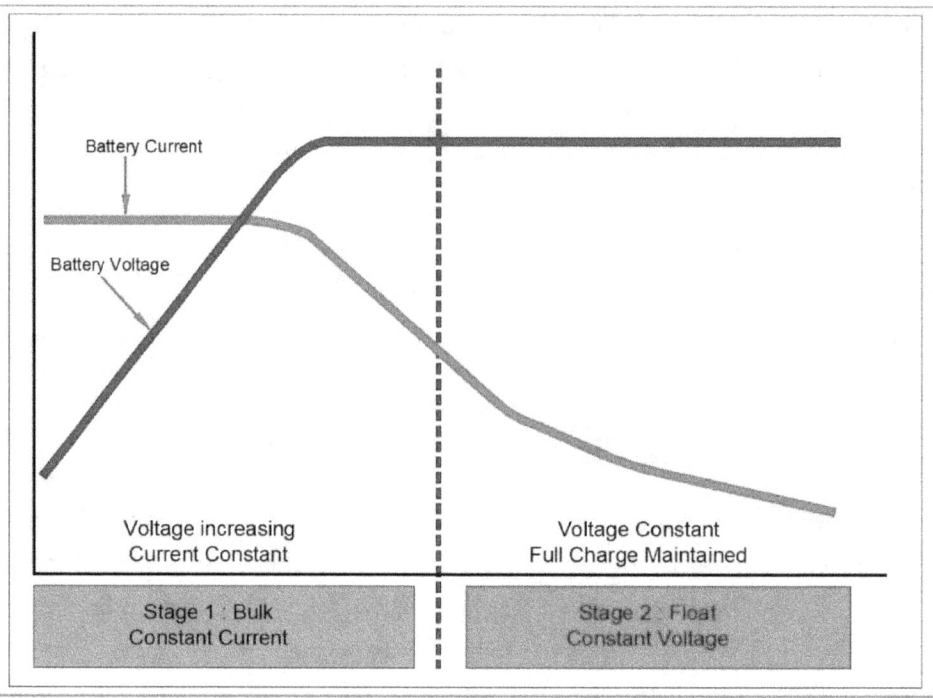

Fig 4.18

2. Constant Voltage (CV) Stage:

Once the peak voltage is reached, the charger transitions to the constant voltage stage. During this stage, the charger maintains a constant voltage while the charge current gradually decreases. The battery's state of charge (SoC) continues to increase during this stage, but at a slower rate. The constant voltage stage is complete when the charging current drops below a set threshold, indicating that the battery is fully charged.

3. Balancing Stage (if applicable):

Some battery management systems include a balancing stage to ensure that all cells in the battery are equally charged. This stage might involve either passive or active balancing. Passive balancing typically involves bleeding off excess charge from cells that are more fully charged, while active balancing involves transferring charge from more charged cells to less charged ones. Not all charging processes for LFP batteries include a balancing stage.

While the charging process for LFP batteries is similar to that for other types of lithium-ion batteries, one key difference is that LFP batteries typically do not require an equalization stage. This is because lithium-ion cells generally do not suffer from the same types of imbalances that can affect lead-acid batteries.

4.9 Effect of Temperature on Battery

As we know, all chemical reactions are affected by temperature, and a battery relies on chemical reaction to generate power. So temperature has important effects on the charge and discharge performance, longevity, and safety of batteries.

Under low-temperature conditions, both Lead acid and Lithium batteries tend to have reduced efficiency. For lead-acid batteries, this is due to the slowing down of internal chemical reactions, which results in diminished capacity and charge acceptance. Lithium-ion batteries also suffer from reduced capacity and charging efficiency at low temperatures; this can result in lithium plating on the anode during charging, a situation that not only degrades battery performance but may also pose safety risks.

Conversely, high temperatures can accelerate degradation processes within both types of batteries. Lead-acid batteries, under high-temperature conditions, tend to self-discharge more quickly and experience accelerated aging. In some cases, overcharging can lead to overheating, which can potentially be hazardous. Similarly, lithium-ion batteries exposed to high temperatures can degrade more rapidly, with the added risk of inducing a thermal runaway condition. This is a dangerous scenario in which the battery rapidly overheats due to the breakdown of the electrolyte.

To ensure the optimal performance and safety of these batteries, it's generally recommended to store and charge them under moderate, preferably room-temperature conditions. Many modern batteries incorporate battery management systems (BMS), which help regulate charging and discharging rates. These systems often include thermal management capabilities to maintain the battery within safe and optimal operating temperature ranges.

4.10 Series and Parallel Connection

When setting up a battery bank for an off-grid solar system, you can connect multiple batteries in either series, parallel, or a combination of both depending on your system's voltage and capacity requirements. Here is a basic overview of what each type of connection means and how it affects the overall battery bank.

1. Series Connection:

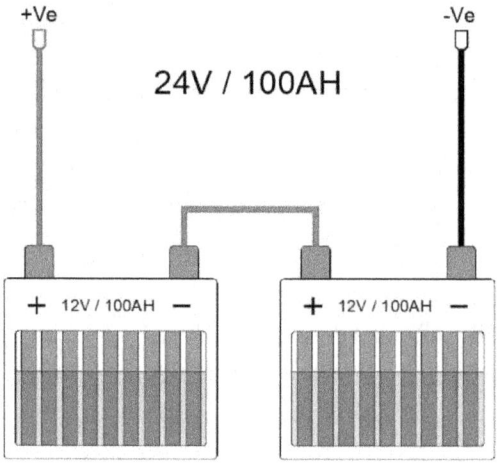

Series Connection

Fig 4.19

When you connect batteries in series, you connect the positive terminal of one battery to the negative terminal of the next. This increases the voltage output while the capacity (measured in ampere-hours, or Ah) remains the same as a single battery.

For example, if you connect two 12-volt (V), 100Ah batteries in series, the total voltage of the battery bank would be 24V (12V + 12V), but the capacity would still be 100Ah.

2. Parallel Connection:

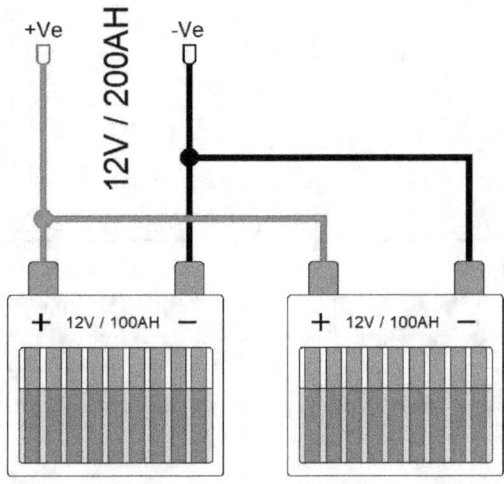

Parallel Connection

Fig 4.20

When you connect batteries in parallel, you connect all the positive terminals together and all the negative terminals together. This keeps the voltage the same while increasing the capacity.

For instance, if you connect two 12V, 100Ah batteries in parallel, the total voltage of the battery bank would still be 12V, but the capacity would increase to 200Ah (100Ah + 100Ah).

3. Series-Parallel Connection:

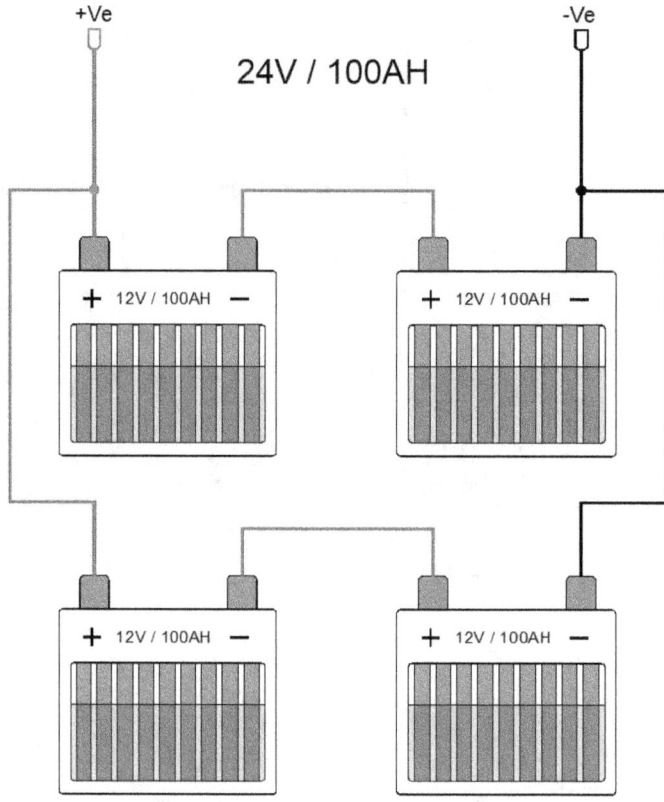

Series Parallel Connection

Fig 4.21

In some cases, you might want to increase both the voltage and capacity of your battery bank. To do this, you can create multiple series or parallel connections and then connect those groups in the opposite arrangement.

For example, you could connect two sets of 12V, 100Ah batteries in series, creating two 24V, 100Ah groups. Then, you could connect those two groups in parallel to create a battery bank with a total voltage of 24V and a total capacity of 200Ah.

Remember, it is crucial to use identical batteries when creating a battery bank (same brand, type, capacity, and age) to ensure balanced charging and discharging. Improper connections can lead to uneven wear, reduced performance, and potential safety risks. Always follow manufacturer guidelines and consider consulting with a professional if you are uncertain.

Chapter 5
Solar Panels

5.1 What is a Solar Panel?

A solar panel, also known as a photovoltaic (PV) module, is a device that converts sunlight into electricity using a process called the photovoltaic effect.

The typical solar panel is composed of individual solar cells, each of which is typically composed of two layers of silicon – one layer is doped with impurities to create an excess of electrons (called an n-type layer), while the other layer is doped to create a deficit of electrons (called a p-type layer). When these two layers are brought together, an electric field is formed at the junction between them, creating a region called the depletion zone.

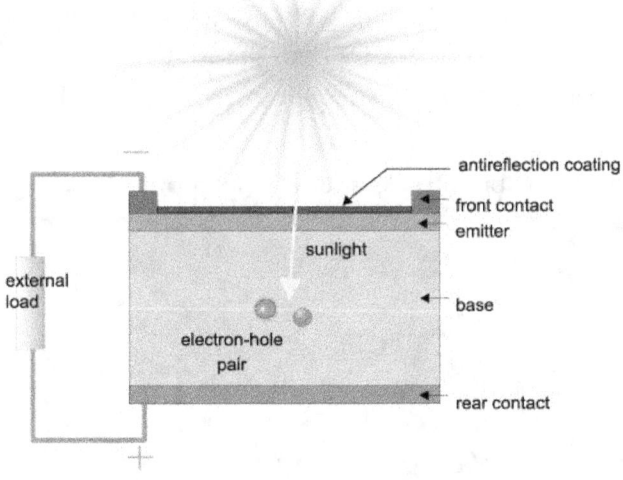

Fig 5.1

Source: pveducation

When sunlight, composed of particles called photons, hits the solar cell, some photons are absorbed by the semiconductor material. The energy from these absorbed photons is transferred to the electrons in the silicon atoms, causing them to break free from their atomic bonds. These free electrons are then pushed by the electric field towards the n-type layer, while the "holes" they leave behind are pushed towards the p-type layer.

By attaching metal contacts to both layers of the solar cell, an external electrical circuit is created. When the free electrons flow through this circuit, an electric current is generated.

5.2 Solar Spectrum Overview

The light spectrum plays a significant role in their efficiency and performance. Sunlight consists of a range of wavelengths that make up the solar spectrum, including ultraviolet (UV) light, visible light, and infrared (IR) light. Each wavelength of light carries a specific amount of energy, with shorter wavelengths (UV) having more energy than longer wavelengths (IR).

Semiconductor materials inside the solar cells can absorb photons across a range of wavelengths. The absorbed photons transfer their energy to the electrons within the semiconductor atoms. The energy required to excite an electron and generate electricity is called the bandgap energy. The efficiency of a solar panel depends on its ability to absorb and convert photons with different energies. Ideally, a solar panel would absorb photons with energies equal to or greater than the bandgap energy. However, photons with higher energy than the bandgap energy can create excess heat, reducing the solar panel's efficiency, while photons with lower energy than the bandgap energy are not absorbed and do not contribute to electricity generation.

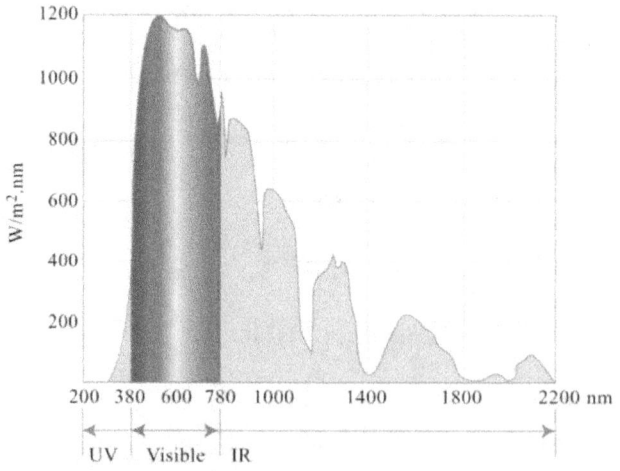

Fig 5.2

Source: MIT OpenCourseWare

Different solar cell materials have different bandgap energies, which affect their ability to absorb and convert light from various parts of the spectrum. For example, silicon solar cells have a bandgap energy that corresponds well with the visible light spectrum, allowing them to efficiently convert this part of the solar spectrum into electricity. However, silicon cells are less efficient at converting UV and IR light. One way to improve the utilization of the light spectrum is by using multijunction solar cells, which are made up of multiple layers of different semiconductor materials, each with a different bandgap energy. This design allows each layer to absorb and convert a specific part of the light spectrum, resulting in higher overall efficiency.

5.3 Solar Irradiance

Solar irradiance refers to the amount of sunlight energy that reaches a specific area on Earth's surface, typically expressed in watts per square meter (W/m^2). It is a vital factor determining the performance and efficiency of solar power systems, such as photovoltaic (PV) panels and solar thermal collectors.

What influences solar irradiance?

Solar irradiance varies due to several factors, such as:

- **Geographical location:** Equatorial regions receive more sunlight than areas near the poles, as the sun's rays are more direct at lower latitudes.
- **Time of day:** Solar irradiance changes throughout the day, with the highest values around solar noon, when the sun is at its zenith.
- **Season:** Irradiance varies with the seasons, due to the Earth's axial tilt, resulting in higher values in the summer and lower values in the winter.
- **Atmospheric conditions:** The atmosphere can absorb, scatter, and reflect sunlight, so factors like cloud cover, air pollution, humidity, and aerosols can also affect the amount of solar radiation that reaches the Earth's surface.

Components of Solar Irradiance

The Direct Normal Irradiance, Global Horizontal Irradiance and Diffuse Horizontal Irradiance are the parameters used for measuring the solar radiation in both sites, following is a definition on the parameters:

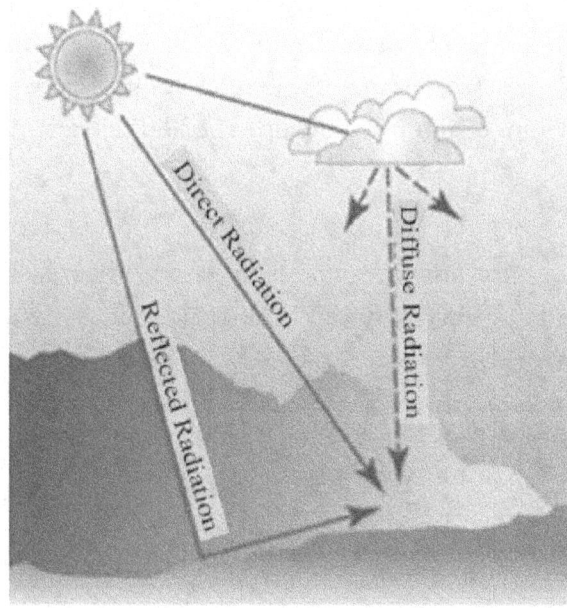

Fig 5.3

Global Horizontal Irradiance (GHI):

GHI represents the total amount of solar radiation received on a horizontal surface, including both direct sunlight and diffuse sunlight scattered by the atmosphere. GHI is an essential parameter for evaluating solar energy potential and designing solar power systems, particularly for flat-mounted solar panels.

Direct Normal Irradiance (DNI):

DNI refers to the solar radiation received directly from the sun on a surface perpendicular to the sun's rays, without any scattering or reflection. DNI is the most concentrated form of solar radiation and is a crucial parameter for designing and evaluating the performance of concentrating solar power (CSP) systems and solar panels with tracking systems.

Diffuse Horizontal Irradiance (DHI):

DHI represents the solar radiation received on a horizontal surface from the sky, excluding direct sunlight, resulting from atmospheric scattering. DHI plays an important role in solar energy applications, especially in regions with significant cloud cover or air pollution, where the diffuse component of solar radiation becomes dominant.

Relationship between GHI, DNI and DHI:

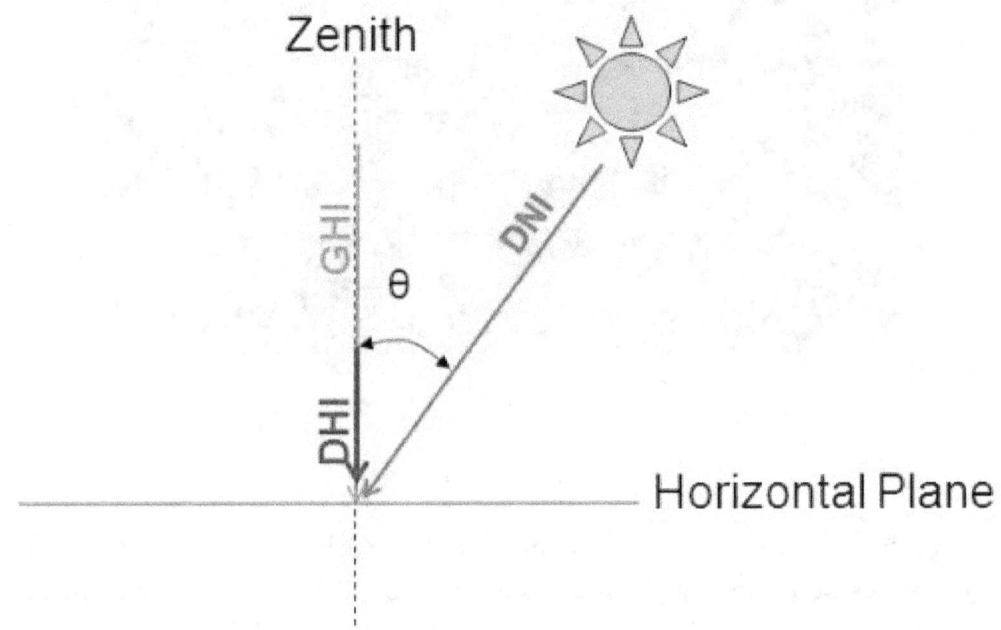

Fig 5.4

The relationship between GHI, DHI, and DNI can be represented by the following equation:

$$GHI = DNI * \cos(\theta) + DHI$$

Where Z is the solar zenith angle (angle between the sun's rays and a line perpendicular to the Earth's surface).

How Irradiance is measured?

Solar Irradiance is measured using specialized instruments called pyranometers and pyrheliometers.

Pyranometers:

Fig 5.5

Source: Rated Power

Pyranometers are instruments used to measure Global Horizontal Irradiance (GHI), which includes both direct and diffuse solar radiation. A pyranometer consists of a thermopile sensor enclosed within a glass or quartz dome. The thermopile sensor absorbs incoming solar radiation, converting it into heat. This temperature difference generates a small voltage proportional to the solar irradiance, which can be recorded and analyzed. The dome serves to protect the sensor while allowing sunlight to pass through and also helps to capture the diffuse solar radiation coming from different directions.

Pyrheliometers:

Fig 5.6

Source: Middleton Solar

Pyrheliometers are used to measure Direct Normal Irradiance (DNI), the solar radiation received directly from the sun on a surface perpendicular to the sun's rays. These instruments consist of a thermopile sensor enclosed within a narrow field-of-view tube with a window at the front. The tube restricts the incoming sunlight to only the direct beam from the sun, excluding diffuse radiation. Like a pyranometer, the thermopile sensor in a pyrheliometer converts the absorbed solar radiation into heat, generating a voltage proportional to the DNI. Pyrheliometers are typically mounted on solar trackers that follow the sun's path, ensuring that the instrument is always aligned with the direct solar beam.

5.4 Radiation Resource Map

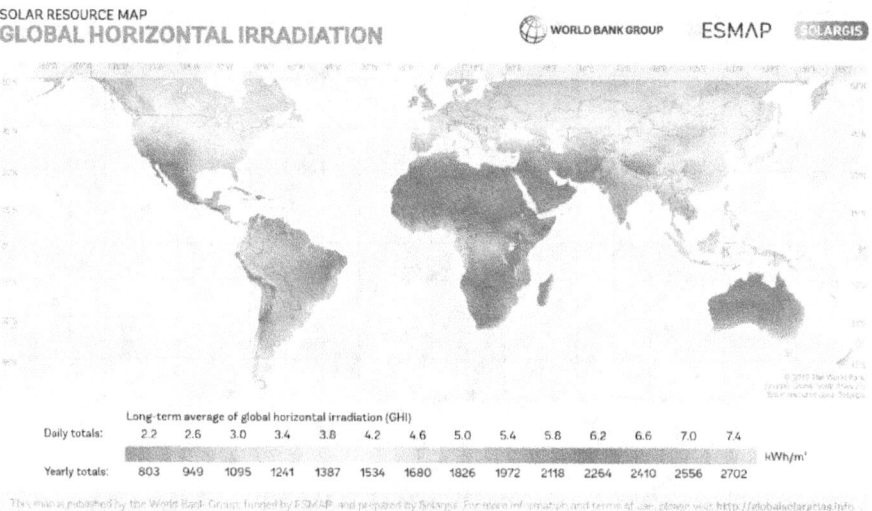

Fig 5.7

Looking at Solar Radiation Resource Map will give you a good idea of the estimated irradiance value in your location. Keep in mind that this is related to GHI, not to direct normal irradiance (DNI).

To see an interactive map visit: https://globalsolaratlas.info/map

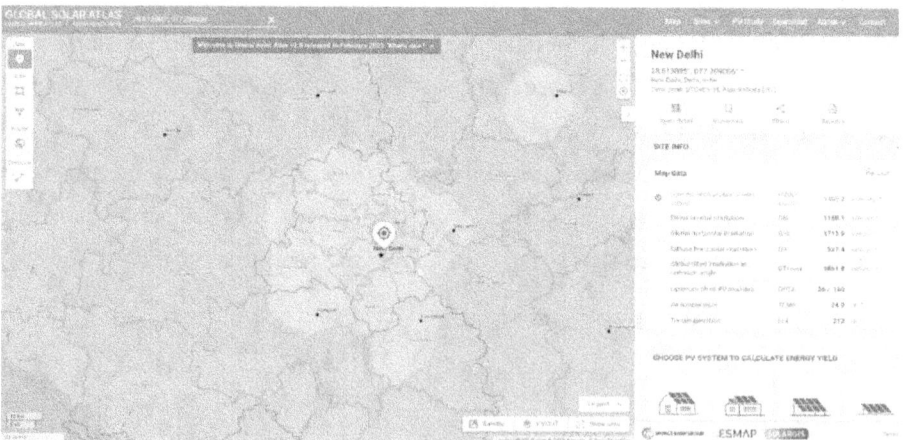

Fig 5.8

If you want to find the solar irradiance of a location using the Global Solar Atlas provided by the World Bank, follow these steps:

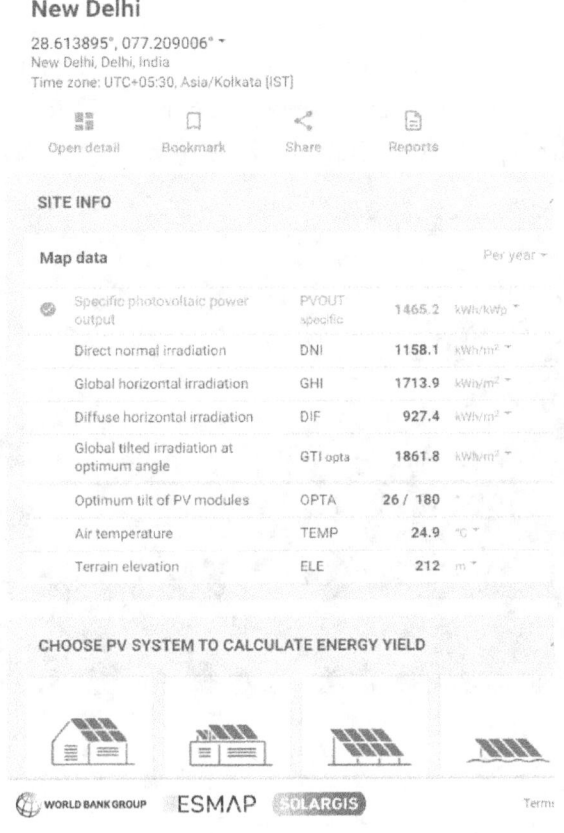

New Delhi

28.613895°, 077.209006° ˅
New Delhi, Delhi, India
Time zone: UTC+05:30, Asia/Kolkata (IST)

Open detail	Bookmark	Share	Reports

SITE INFO

Map data   Per year ˅

	Specific photovoltaic power output	PVOUT specific	1465.2	kWh/kWp ˅
	Direct normal irradiation	DNI	1158.1	kWh/m² ˅
	Global horizontal irradiation	GHI	1713.9	kWh/m² ˅
	Diffuse horizontal irradiation	DIF	927.4	kWh/m² ˅
	Global tilted irradiation at optimum angle	GTI opta	1861.8	kWh/m² ˅
	Optimum tilt of PV modules	OPTA	26 / 180	
	Air temperature	TEMP	24.9	°C ˅
	Terrain elevation	ELE	212	m ˅

CHOOSE PV SYSTEM TO CALCULATE ENERGY YIELD

WORLD BANK GROUP   ESMAP   SOLARGIS   Terms

Fig 5.9

Step-1: Visit the Global Solar Atlas website: Go to the website at https://globalsolaratlas.info.

Step-2: Search for a specific location by entering the address, city, or coordinates in the search bar at the top left corner of the page.

Step-3: Click on your desired location on the map, and a pop-up window will appear, displaying relevant solar irradiance data for that point. This includes annual averages of Global Horizontal Irradiance (GHI) and Direct Normal Irradiance (DNI), as well as the estimated electricity production potential for photovoltaic (PV)

After getting the above radiation data, you can review it for your location and use it to assess the solar PV potential, and estimate the expected energy production.

5.5 Types of Solar Panels

There are several types of solar panels available in the market, each with its unique characteristics, advantages, and disadvantages. There are three primary types of solar panels: monocrystalline, polycrystalline, and thin-film.

1. Monocrystalline solar panels:

Fig 5.10

Source: climatebiz

These panels are made from a single, continuous crystal structure of high-purity silicon. Monocrystalline panels are characterized by their uniform, dark black appearance and rounded edges. They offer the highest efficiency among all solar panel types, typically ranging from 18% to 24%. However, they are also generally more expensive due to their manufacturing process and the high quality of silicon used.

2. Polycrystalline solar panels:

Fig 5.11

Source: pitech

Polycrystalline panels are made from multiple silicon crystals that are melted and poured into a mold. They have a speckled, blue appearance due to the presence of multiple crystal structures. Polycrystalline solar panels have slightly lower efficiency than monocrystalline panels, typically ranging from 15% to 20%. However, they are generally more affordable, making them a popular choice for residential and commercial installations.

3. Thin-film solar panels:

Fig 5.12

Source: solarpowerworld

Thin-film panels are made by depositing a thin layer of semiconductor material, such as amorphous silicon (a-Si), cadmium telluride (CdTe), or copper indium gallium selenide (CIGS), onto a substrate like glass, metal, or plastic. These panels are characterized by their flexibility, lightweight, and uniform appearance. Thin-film solar panels have lower efficiency than crystalline panels, usually ranging from 10% to 12%. However, they can be more flexible and lightweight, making them suitable for specific applications like building-integrated photovoltaics (BIPV) or portable solar chargers. Thin-film panels are also less affected by high temperatures and shading compared to crystalline panels.

5.6 New Solar Panel Technologies in the Market

In addition to the above 3 types of solar panels, advanced technologies and designs, such as PERC, bifacial, and half-cut cells, have been developed to enhance solar panel performance. These technologies can be applied to both monocrystalline and polycrystalline panels.

1. **PERC (Passivated Emitter and Rear Cell) technology:**

Fig 5.13

Source: Solar Sam

PERC is an advanced solar cell technology that improves the efficiency of solar panels. In PERC solar cells, a passivation layer is added to the rear surface of the solar cell, which helps to reflect light that would otherwise be lost back into the cell for additional absorption. This increases the overall efficiency of the solar cell. PERC technology can be found in both monocrystalline and polycrystalline solar panels.

2. **Bifacial solar panels:**

Fig 5.14

Source: saurenergy

Bifacial panels are designed to capture sunlight from both the front and back sides of the panel, increasing their overall energy production. These panels typically use monocrystalline or polycrystalline solar cells and have a transparent back sheet or glass that allows light to pass through and reflect off the surface behind the panel. Bifacial panels can achieve higher efficiency and are particularly useful in installations where the backside of the panel is exposed to sunlight, such as on tracking systems or elevated structures.

3. Half-cut cells:

Fig 5.15

Source: SolarLabs

Half-cut solar panels consist of solar cells that have been cut in half, reducing resistive losses and increasing overall panel efficiency. They have a higher tolerance for shading and can maintain better performance under partial shading conditions compared to traditional full-cell solar panels. This technology is advantageous for off-grid solar systems located in areas with variable sunlight conditions or where shading is a concern.

5.7 Emerging Solar Panel Technologies

Solar panel technologies are constantly being developed to improve efficiency, reduce costs, and increase the range of applications for solar energy. Some of the most promising emerging solar panel technologies include:

1. Heterojunction (HJT) Solar Cells

    Heterojunction solar cells combine thin layers of amorphous silicon (a-Si) with crystalline silicon (c-Si) to create a highly efficient solar cell. HJT cells benefit from the excellent light absorption properties of amorphous silicon and the high carrier mobility of crystalline silicon. These cells can achieve efficiency rates above 23% and offer better temperature performance compared to traditional silicon solar cells.

2. **N-Type TOPCon Solar Cells**

N-Type Tunnel Oxide Passivated Contact (TOPCon) solar cells are an emerging technology that utilizes n-type silicon wafers with passivated contacts on both the front and rear sides of the cell. This design reduces electron recombination and improves carrier collection, leading to higher efficiency rates. N-Type TOPCon solar cells have demonstrated efficiencies above 24% in laboratory settings and have the potential to become a competitive alternative to conventional solar cells in the future.

3. **Perovskite Solar Cells**

Perovskite solar cells use a unique crystalline structure made from a combination of organic and inorganic materials. These cells have the potential to achieve higher efficiency rates than traditional silicon-based solar cells, with current laboratory efficiencies reaching over 25%. Perovskite solar cells can also be manufactured using cost-effective techniques, such as inkjet printing, which could significantly reduce the cost of solar energy production.

4. **Tandem Solar Cells**

Tandem solar cells combine two or more layers of photovoltaic materials with different bandgaps to capture a broader range of the solar spectrum. By stacking these layers, tandem cells can achieve higher efficiency rates than traditional single-junction solar cells. For example, perovskite-silicon tandem cells have demonstrated efficiencies above 29% in laboratory settings, making them an attractive option for future solar power systems.

5.8 Conversion Efficiency

The conversion efficiency of a solar panel is a measure of how effectively the panel converts sunlight into electrical energy. It is expressed as a percentage and can be calculated using the following formula:

Solar Panel Efficiency (%) = (Electrical Power Output / Incident Solar Power) x 100

Where:

- Electrical Power Output is the amount of electrical power produced by the solar panel, typically measured in watts (W) or kilowatts (kW).

- Incident Solar Power is the amount of sunlight striking the panel's surface, typically measured in watts per square meter (W/m²) or kilowatts per square meter (kW/m²).

5.9 Characteristics Curves

I-V Curve:

The I-V curve is a graphical representation of the relationship between the current (I) and voltage (V) produced by a solar panel under varying illumination and temperature conditions. The I-V curve provides valuable insights into the performance of a solar panel and helps in identifying its operating points, such as the maximum power point (MPP), open-circuit voltage (Voc), and short-circuit current (Isc).

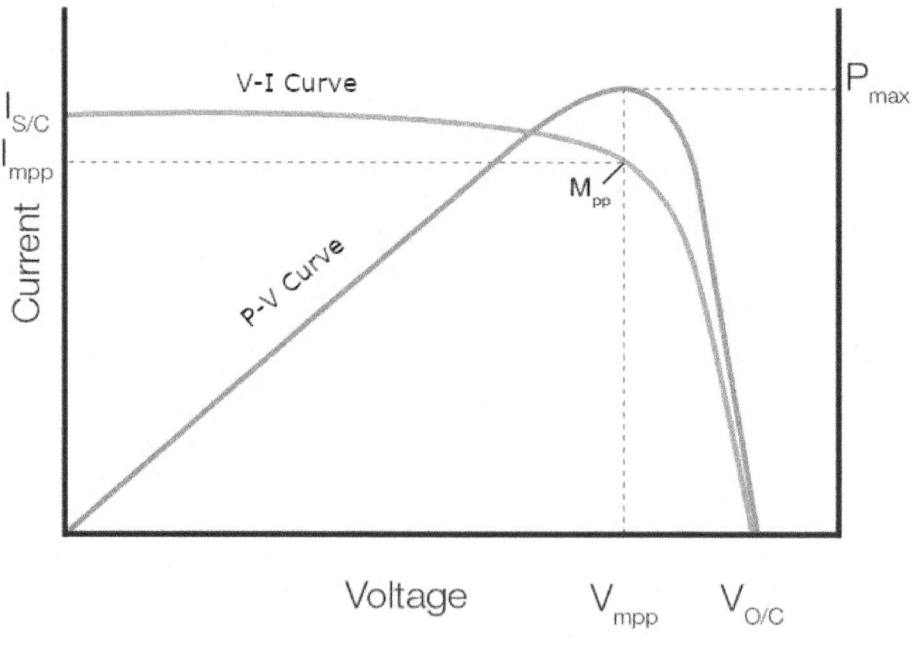

Fig 5.16

1. Operating point:

The point on the VI curve that corresponds to the solar panel's operating condition (i.e., the voltage and current at which the panel is producing power) is called the operating point.

2. Open-Circuit Voltage (Voc):

This is the maximum voltage produced by the solar panel when it is not connected to any load, i.e., when the current is zero. At this point, the solar panel generates the highest voltage possible under given illumination and temperature conditions. Voc is useful for understanding the panel's voltage limits and selecting compatible system components, such as charge controllers and inverters.

3. Short-Circuit Current (Isc):

This is the maximum current produced by the solar panel when it is short-circuited, i.e., when the voltage across the panel is zero. Isc indicates the panel's ability to generate current under specific illumination and temperature conditions. It is an important parameter for designing protective devices, such as fuses or circuit breakers, and determining the required wire sizes in a solar system.

4. Maximum Power Point (MPP):

The MPP represents the operating point on the I-V curve where the solar panel produces the maximum power output. At this point, the product of the current and voltage is the highest, resulting in the greatest energy production. The MPP is influenced by factors such as temperature and irradiance and is critical for optimizing solar panel performance. Maximum power point tracking (MPPT) charge controllers are designed to continually adjust the panel's operating point to maintain the MPP for maximum energy production.

5. Fill Factor (FF):

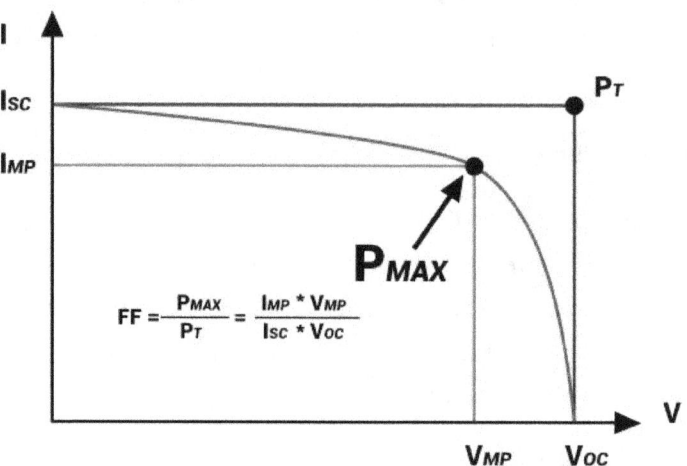

Fig 5.17

The fill factor is a measure of the solar panel's efficiency and is calculated by dividing the maximum power output by the product of Voc and Isc. It provides an indication of how closely the I-V curve approaches a perfect rectangle. A higher fill factor signifies better solar cell quality and more efficient energy conversion.

Here is how to interpret an IV curve:

The x-axis represents the output voltage, ranging from 0V to the open-circuit voltage (Voc) of the solar panel. The y-axis represents the output current, ranging from 0A to the short-circuit current (Isc) of the solar panel.

The IV curve has three main regions:

- **Short-circuit region (0V to Vmp):** In this region, as the voltage increases from 0V, the current starts to decrease from the short-circuit current (Isc). The curve is steep initially and then starts to flatten as the voltage approaches the maximum power point (Vmp).

- **Maximum power point (Vmp, Imp):** This is the point on the curve where the solar panel generates the maximum power. The voltage and current at this point are called the rated voltage (Vmp) and rated current (Imp), respectively.

- **Open-circuit region (Vmp to Voc):** In this region, as the voltage increases from Vmp towards the open-circuit voltage (Voc), the current drops sharply from Imp to 0A.

The point on the IV curve where the power output is maximum is called the Maximum Power Point (MPP). The MPP represents the operating condition where the solar panel is most efficient and delivers the highest power output.

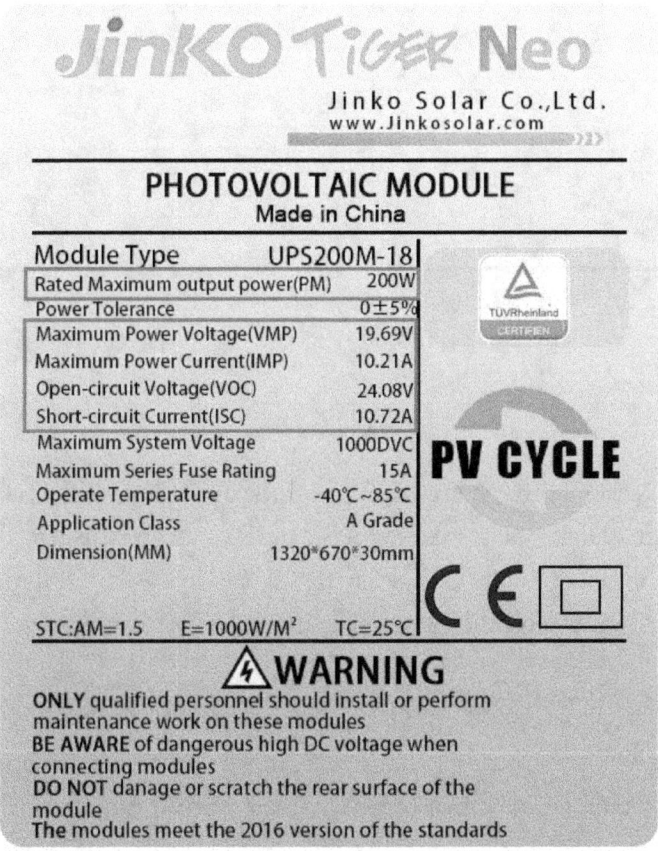

Fig 5.18

Source: Jinko Solar

From the above datasheet of 200W panel, the key parameters are:

- Maximum Power (Pmax): 200W
- Maximum Power Voltage (Vmp): 19.69V
- Maximum Power Current (Imp): 10.21A
- Open-circuit Voltage (Voc): 24.08V
- Short-circuit Current (Isc): 10.72A

P-V Curve:

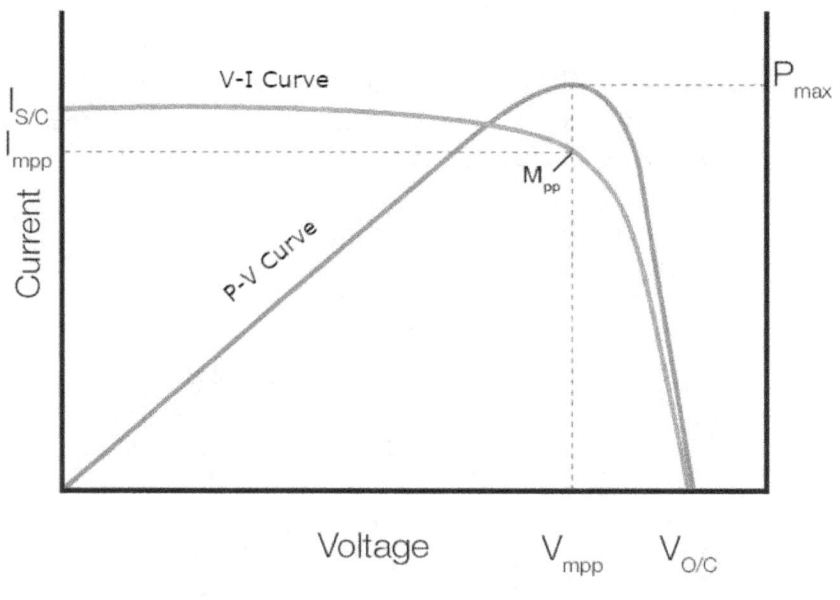

Fig 5.19

A PV curve, or Power-Voltage curve, represents the relationship between the output power and output voltage of a solar panel. The PV curve is useful for understanding the performance and efficiency of a solar panel, as well as determining the maximum power point (MPP) at which the panel operates most effectively.

Here's how to interpret a PV curve:

The x-axis represents the output voltage, ranging from 0V to the open-circuit voltage (Voc) of the solar panel. The y-axis represents the output power, which is the product of the output voltage and output current (Power = Voltage × Current).

The PV curve has two main regions:

- **Rising region (0V to MPP):** In this region, the output power increases as the output voltage increases. The curve starts at the origin (0V, 0W) and rises as the voltage increases. This is because the current remains relatively high while the voltage increases, resulting in a higher power output.

- **Falling region (MPP to Voc):** In this region, the output power decreases as the output voltage increases. Beyond the MPP, the increase in voltage is accompanied by a sharp decrease in current, causing the power output to drop.

The point on the PV curve where the power output is maximum is called the Maximum Power Point (MPP). The voltage and current at the MPP are referred to as the rated voltage (Vmp) and rated current (Imp), respectively. The MPP represents the operating condition where the solar panel is most efficient and delivers the highest power output.

Effect of Insolation and Temperature:

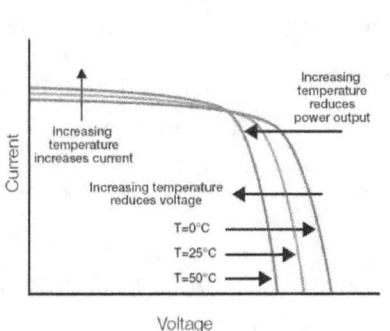 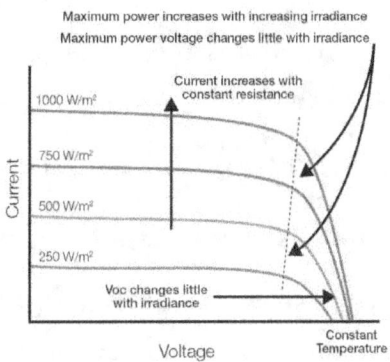

Fig 5.20

Source: seaward

Insolation and temperature significantly affect the shape and position of these curves.

Effect of Insolation:

When insolation increases, the amount of sunlight absorbed by the solar panel also increases. This leads to a higher output current (I) at any given voltage (V). As a result, the I-V curve shifts upward, and the maximum power point (MPP) on the P-V curve moves to a higher power output level. Conversely, when insolation decreases, the output current decreases, the I-V curve shifts downward, and the MPP on the P-V curve moves to a lower power output level.

Effect of Temperature:

Temperature affects both the output current and voltage of solar panels but in different ways.

- **Output current (I):** As temperature increases, the output current of a solar panel typically increases slightly. However, this increase in current is relatively small compared to the change in voltage due to temperature.

- **Output voltage (V):** As temperature increases, the output voltage of a solar panel decreases. Solar cells are made of semiconductor materials, and their voltage output is sensitive to temperature. The voltage decrease is more significant than the current increase, which results in an overall reduction in power output.

When the temperature rises, the I-V curve shifts to the left, and the MPP on the P-V curve moves to a lower voltage and power output level. Conversely, when the temperature decreases, the I-V curve shifts to the right, and the MPP on the P-V curve moves to a higher voltage and power output level.

5.10 STC and NOCT

STC (Standard Test Conditions):

STC stands for "Standard Test Conditions" and are the industry standard for the conditions under which a solar panel are tested. By using a fixed set of conditions, all solar panels can be more accurately compared and rated against each other.

Think of it as a controlled environment where all solar panels are tested under the same conditions to make it easier for us to compare their performance. The conditions used in STC are:

- **Temperature - 25°C:** The temperature of the solar cell itself, not the temperature of the surrounding.

- **Sunlight Intensity (Irradiance)-1000 W/m²:** This number refers to the amount of light energy falling on a given area at a given time.

- **Air Mass (AM) - 1.5:** Air mass (AM) is a measure of the path length that sunlight travels through the Earth's atmosphere before reaching a specific point on the ground. As the sun moves across the sky, the angle at which sunlight enters the atmosphere changes, affecting the distance it has to travel through the atmosphere.

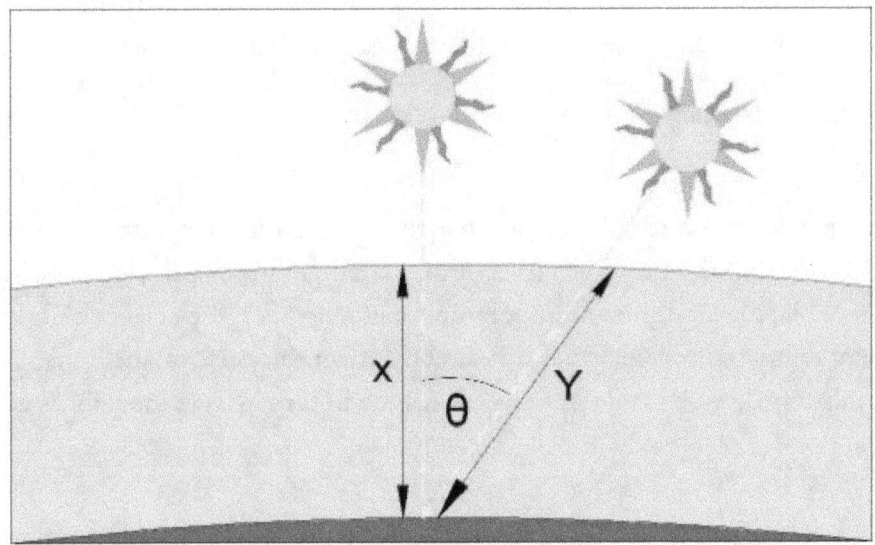

Fig 5.21

Source: silicon solar

Air mass 1.5 (AM1.5) is a standard measurement used in the solar industry to represent a typical sunlight condition when the sun is at an angle of around 48 degrees from the horizon. At this angle, sunlight has to travel through 1.5 times the amount of atmosphere compared to when the sun is directly overhead (AM1.0).

NOCT (Nominal Operating Cell Temperature):

While STC provides a good starting point for comparing solar panels, it does not account for real-world conditions that solar panels will actually face. That is where NOCT comes in.

NOCT is an estimate of the temperature at which the solar cells will actually operate when exposed to real-world sunlight, wind, and outdoor temperatures. Solar panels tend to be less efficient as they get hotter, which means that their power output might be lower than what we would expect based on their STC rating.

In simple terms, NOCT gives us a more realistic idea of how a solar panel will perform when it is installed on your roof or in a solar farm, as it takes into account the higher temperatures that the solar cells will experience in real-world conditions.

5.11 Ambient Temperature and Cell Temperature

Ambient temperature and cell temperature are two different but related factors that affect the performance of solar panels. Let us discuss each in detail:

Ambient Temperature:

Ambient temperature refers to the temperature of the surrounding air. It is an essential factor to consider when installing solar panels, as it can influence the temperature of the solar cells within the panels. Higher ambient temperatures can cause the solar cells to heat up, while lower ambient temperatures can help keep the cells cooler. However, the ambient temperature itself does not directly affect the performance of solar panels; rather, it impacts the solar cell temperature, which in turn affects the panel's efficiency.

Cell Temperature:

Cell temperature refers to the actual temperature of the solar cells within a solar panel. The cell temperature is influenced by several factors, including ambient temperature, sunlight intensity (insolation), and the panel's ability to dissipate heat. When solar cells absorb sunlight, they convert some of the energy into electricity and some into heat. This heat raises the cell temperature, which can negatively impact the solar panel's efficiency.

Cell temperature plays a crucial role in solar panel performance because the efficiency of solar cells decreases as their temperature increases. For instance, a solar panel's output voltage decreases with increasing cell temperature, resulting in a lower power output. To account for the effects of cell temperature on solar panel performance, the solar industry uses the Nominal Operating Cell Temperature (NOCT) parameter. NOCT is a standardized estimate of the cell temperature under typical real-world operating conditions, including sunlight, wind, and ambient temperature.

5.12 Temperature coefficients

Temperature coefficients are essential parameters that describe how the performance of solar panels changes in response to changes in temperature. There are three primary temperature coefficients associated with solar panels: the temperature coefficient of power (Pmax), the temperature coefficient of voltage (Voc), and the temperature coefficient of current (Isc).

Temperature Coefficient of Power (Pmax):

This coefficient describes the percentage change in the maximum power output (Pmax) of a solar panel for every degree Celsius change in temperature. It is usually expressed as a percentage per degree Celsius (%/°C). A typical value for the temperature coefficient of power is -0.4%/°C for crystalline silicon solar panels, meaning that the power output decreases by 0.4% for every 1°C increase in temperature.

Temperature Coefficient of Voltage (Voc):

This coefficient describes the change in the open-circuit voltage (Voc) of a solar panel for every degree Celsius change in temperature. It is usually expressed in volts per degree Celsius (V/°C). The voltage of solar panels typically decreases with increasing temperature. For example, a temperature coefficient of voltage for a crystalline silicon solar panel might be around -0.3 V/°C, meaning that the open-circuit voltage decreases by 0.3 volts for every 1°C increase in temperature.

Temperature Coefficient of Current (Isc):

This coefficient describes the change in the short-circuit current (Isc) of a solar panel for every degree Celsius change in temperature. It is usually expressed in amps per degree Celsius (A/°C). The current of solar panels typically increases slightly with increasing temperature. For instance, a temperature coefficient of current for a crystalline silicon solar panel might be around 0.045 A/°C, meaning that the short-circuit current increases by 0.045 amps for every 1°C increase in temperature.

Example Calculation:

Now let us see an actual datasheet of a solar panel and calculate the characteristics values at 35°C

SPECIFICATIONS						
Module Type	JKM530M-72HL4 JKM530M-72HL4-V		JKM535M-72HL4 JKM535M-72HL4-V		JKM540M-72HL4 JKM540M-72HL4-V	
	STC	NOCT	STC	NOCT	STC	NOCT
Maximum Power (Pmax)	530Wp	394Wp	535Wp	398Wp	540Wp	402Wp
Maximum Power Voltage (Vmp)	40.56V	37.84V	40.63V	37.91V	40.70V	38.08V
Maximum Power Current (Imp)	13.07A	10.42A	13.17A	10.50A	13.27A	10.55A
Open-circuit Voltage (Voc)	49.26V	46.50V	49.34V	46.57V	49.42V	46.65V
Short-circuit Current (Isc)	13.71A	11.07A	13.79A	11.14A	13.85A	11.19A
Module Efficiency STC (%)	20.55%		20.75%		20.94%	
Operating Temperature(°C)					-40°C~+85°C	
Maximum system voltage					1000/1500VDC (IEC)	
Maximum series fuse rating					25A	
Power tolerance					0~+3%	
Temperature coefficients of Pmax					-0.35%/°C	
Temperature coefficients of Voc					-0.28%/°C	
Temperature coefficients of Isc					0.048%/°C	
Nominal operating cell temperature (NOCT)					45±2°C	

Fig 5.22

Source: Jinko Solar

From the above datasheet, following are parameters under STC at 25°C:

- Maximum power output (Pmp): 540 W
- Open-circuit voltage (Voc): 40.7 V
- Short-circuit current (Isc): 13.85 A

Temperature coefficients are as follows:

- Power (Pmax): -0.35%/°C
- Voltage (Voc): -0.28 V/°C
- Current (Isc): -0.048 A/°C

If the temperature increases to 35°C, the change in temperature is (35 - 25) = 10°C. Now, let us calculate the new values of power, voltage, and current at 35°C:

Power (Pmp): With a temperature coefficient of -0.35%/°C and a 10°C increase, the power output would decrease by 3.5% (0.35% x 10). So, the decrease in power output is 540 W * 3.5% = 18.9 W.

The new power output at 35°C is 540 W - 18.9 W = 521.1 W.

Voltage (Voc): With a temperature coefficient of -0.28 V/°C and a 10°C increase, the open-circuit voltage would decrease by 2.8 V (0.28 V/°C x 10).

The new open-circuit voltage at 35°C is 40.7 V - 2.8 V = 37.9 V.

Current (Isc): With a temperature coefficient of -0.048 A/°C and a 10°C increase, the short-circuit current would increase by 0.48 A (-0.048 A/°C x 10).

The new short-circuit current at 35°C is 13.85 A + 0.48 A = 14.33 A.

At 35°C, the new values for the solar panel are:

- Maximum power output (Pmp): **521.1 W**
- Open-circuit voltage (Voc): **37.9 V**
- Short-circuit current (Isc): **14.33 A**

5.13 Series and Parallel Connections

In solar power systems, there are three common methods for connecting solar panels: series, parallel, and series-parallel (also called mixed or combined connection). Each connection type has its unique characteristics, advantages, and disadvantages.

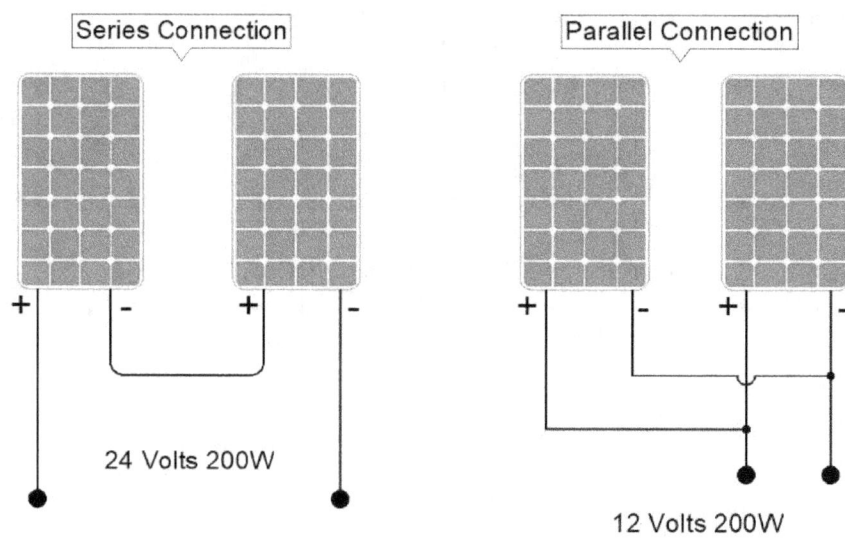

Fig 5.23

Series Connection:

In a series connection, solar panels are connected end-to-end in a single path, allowing the current to flow through each panel sequentially. The solar panels connected in series is known as a string. The total voltage across the panels is the sum of their individual voltages, while the current remains constant.

Example: Let us consider a solar panel rated at 100W and 12V, two solar panels are connected in series. The voltage of each panel is added, while the current remains the same as a single panel.

Total Voltage = 12V + 12V = 24V

Total Power = 100W+100W = 200W

Parallel Connection:

In a parallel connection, solar panels are connected with their positive terminals together and their negative terminals together, forming multiple current paths. The total current is the sum of their individual currents, while the voltage remains constant.

Example:

Total Voltage = same as a single panel (12V)

Total Power = 100W+100W = 200W

Series-Parallel Connection:

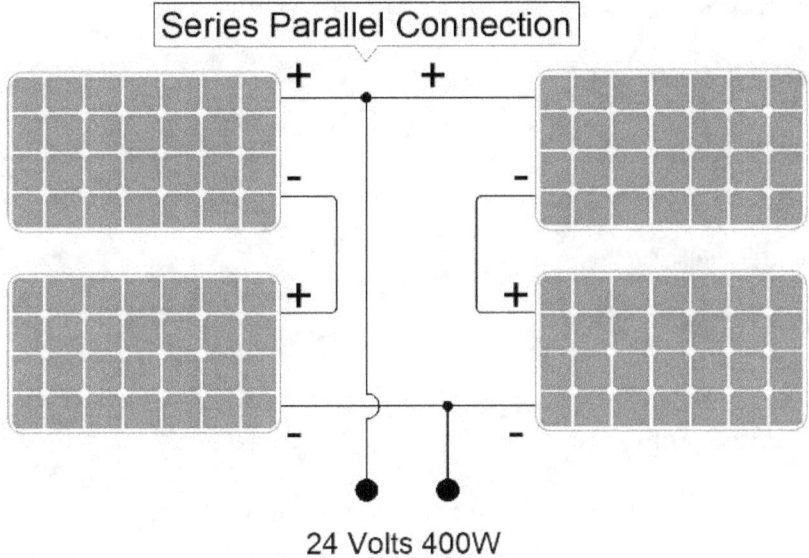

24 Volts 400W

Fig 5.24

In a series-parallel connection, solar panels are arranged in a combination of both series and parallel connections, which is also known as solar panel array. This method is commonly used in solar power systems to achieve the desired voltage and current levels.

For example, two same solar panels are connected in series to form strings, and these strings are then connected in parallel.

Total Voltage = Voltage (String-1) + Voltage (String -2) = 12V + 12V = 24V

Total Power = 200W+200W = 400W

5.14 Combining Different Solar Panel Types

In some cases, you may want or need to use different types of solar panels in your off-grid solar system. Combining different solar panels can be beneficial for maximizing your system's efficiency and performance. However, it's essential to consider the compatibility of various solar panels and how to connect them properly. In this section, we will discuss the factors to consider when combining different solar panel types and provide an example to illustrate how to connect them effectively.

Example:

Suppose you have two 100W monocrystalline solar panels and two 100W polycrystalline solar panels for your off-grid solar system. The specifications of each panel are as follows:

Monocrystalline Panels:

- Rated Voltage (Vmp): 20V
- Rated Current (Imp): 5A

Polycrystalline Panels:

- Rated Voltage (Vmp): 18V
- Rated Current (Imp): 5.6A

Electrical Compatibility:

The voltage and current ratings of the two solar panel types are relatively close, which is essential for ensuring compatibility. Although not identical, the differences in Vmp and Imp are small enough that the panels can be combined with minimal power losses and risks to the solar panels or other system components.

Connection Strategy:

Since the solar panels have similar, but not identical, voltage and current ratings, connecting them in series may not be the best option. Instead, connect the panels of the same type in series first, creating two strings: one with the monocrystalline panels and another with the polycrystalline panels. Then, connect these two strings in parallel to minimize the impact of the differences in voltage and current ratings on system efficiency.

Monocrystalline String:

- Total Voltage (Vmp): 20V + 20V = 40V
- Total Current (Imp): 5A

Polycrystalline String:

- Total Voltage (Vmp): 18V + 18V = 36V
- Total Current (Imp): 5.6A

Parallel Connection of Strings:

- Total Voltage (Vmp): Approximately 38V (average of 40V and 36V)
- Total Current (Imp): 5A + 5.6A = 10.6A

5.15 Tilt & Azimuth Angle

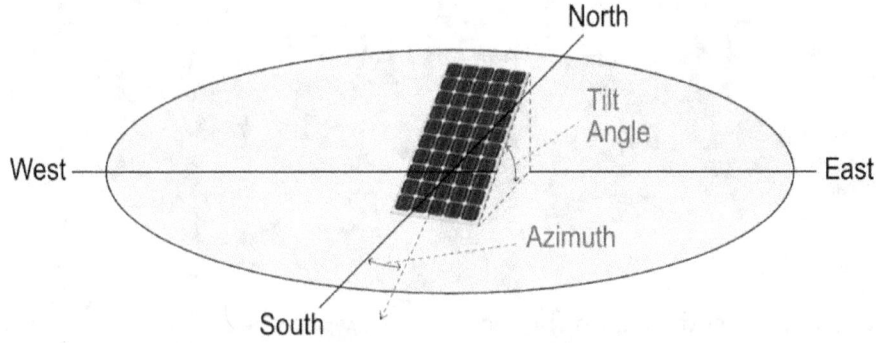

Fig 5.25

Source: Solar Design Guide

Tilt and azimuth angles are essential factors in designing and installing solar energy systems. These angles determine the orientation of solar panels or solar arrays to maximize the amount of sunlight they receive, leading to increased efficiency and overall power generation.

Tilt Angle:

The tilt angle refers to the angle between a solar panel's surface and the ground or horizontal plane. The optimal tilt angle for a solar panel depends on the location's latitude and the time of year.

- For fixed solar panel installations, the tilt angle is often set to equal the location's latitude to maximize annual energy production. However, the optimal tilt angle can be adjusted depending on seasonal variations.

- For maximum efficiency in the winter, the tilt angle can be increased by around 10 degrees from the latitude, while in summer, it can be decreased by around 10 degrees. This compensates for the sun's changing position in the sky throughout the year.

- Some solar installations use tracking systems that automatically adjust the tilt angle to follow the sun's path across the sky, maximizing daily energy production.

Here is how you can find out the latitude of your location:

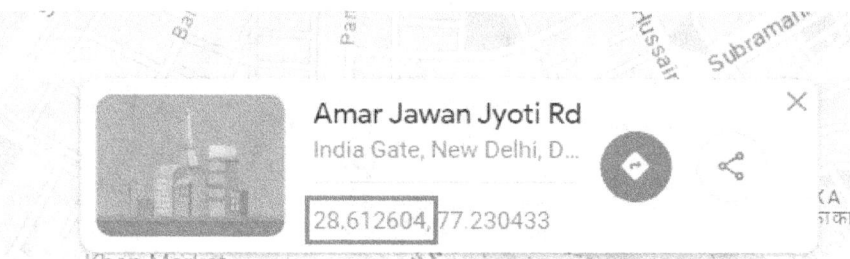

Fig 5.26

Go to google maps and click on the location you would like to know the latitude. The first number is the latitude. The one after it is the longitude. In this example, the latitude is 28°.

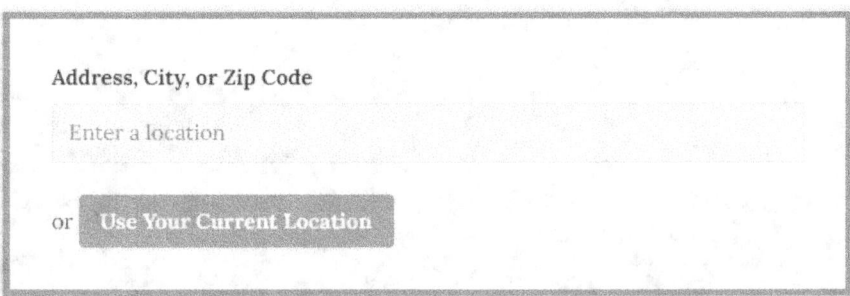

Fig 5.27

Online solar calculators or tools can help determine the best orientation based on your location's latitude and other factors.

You can use this online calculator to determine the best tilt angle for your location.

Azimuth Angle:

The azimuth angle refers to the horizontal angle measured between a reference direction (usually true north) and the direction a solar panel faces.

- In the Northern Hemisphere, solar panels are typically oriented towards true south to maximize sunlight exposure, while in the Southern Hemisphere, they should be oriented towards true north.

- The exact azimuth angle for maximum efficiency can vary depending on the specific location and time of year.

5.16 Shading

Solar panel shading is a significant concern in solar energy systems, as it can greatly reduce the efficiency and overall power generation of the system. There are mainly two types of shading: near-shading and far-shading.

Fig 5.28

Source: SolarLabs

Near Shadows:

Near shadows are caused by objects in close proximity to the solar panels, such as trees, buildings, or even other solar panels. These objects can cast shadows on the solar panels, reducing their exposure to sunlight and lowering their power output. Near shadows can have a significant impact on the performance of the solar energy system, especially when the objects causing the shadows are large or tall.

Far Shadows:

Far shadows are caused by objects located at a greater distance from the solar panels. While these objects may be farther away, their size, height, or position relative to the sun can still result in shadows on the solar panels. Far shadows generally have a less severe impact on the solar panels' performance compared to near shadows, as they tend to cover a smaller area of the panel and may result in partial shading rather than complete shading.

The effects of shading on solar panels can be severe, including:

- Reduced power output: When a solar panel is shaded, it produces less power due to decreased exposure to sunlight. This directly affects the overall power generation of the solar energy system, resulting in lower energy production and reduced efficiency.

- Hot spots: Shading can cause a phenomenon called "hot spots," where the shaded solar cells experience increased heat due to the flow of current from the illuminated cells. This can cause damage to the solar panel and reduce its lifespan.

- Impact on series-connected panels: In a series-connected solar panel array, the performance of the entire string is limited by the weakest panel. If one panel is shaded, it can severely impact the current and power output of the entire string, resulting in significant power loss.

How to Minimize the Shadow Effects?

To minimize the effects of both near and far shadows, consider the following factors during the design and installation of solar energy systems:

- Choose a location with minimal shading from nearby objects, such as trees or buildings.

- Properly orient and space the solar panels to minimize the potential for shading. This may involve adjusting the tilt and azimuth angles or increasing the distance between panels to prevent shadows from one panel falling on another.

- Use bypass diodes, microinverters or power optimizers to mitigate the effects of shading on the solar panels' performance and prevent issues such as hot spots or reduced power output.

5.17 Solar Panel Hotspot

A hotspot refers to a localized area within a solar panel that becomes significantly hotter than the surrounding areas. Hotspots occur when one or more solar cells within the panel are shaded, damaged, or have manufacturing defects, causing them to generate less power or even act as resistors instead of power generators.

A hotspot in a solar panel is like a traffic jam on a road. Imagine a road with multiple lanes where all the cars (electric current) need to move together. Now, if there is an accident or a broken-down car in one of the lanes (a shaded or damaged solar cell), it causes a traffic jam (hotspot), slowing down all the cars behind it and creating congestion. The traffic jam (hotspot) generates heat and can damage the road (solar panel).

When multiple solar cells are connected in series, the overall operating current is determined by the weakest (or bad) cell in the string. If one cell is shaded, its short-circuit current decreases, limiting the current flowing through the entire string. When the other cells in the series string become forward biased due to power generation, the shaded cell becomes reverse biased. As a result, the entire power generated by the other cells in the string is dissipated as heat in the shaded cell.

Another factor that contributes to hot spots is dust and debris build-up. Dirt and dust on the surface of the panel can block some of the sunlight that would normally reach the cells, leading to a reduction in performance and an increase in temperature.

Fig 5.29

How to prevent hot spots

Let us continue with our earlier traffic jam example, To avoid this problem, an alternative route (bypass diode) is provided, which allows the cars (electric current) to go around the traffic jam (hotspot), keeping the traffic moving smoothly and preventing further damage to the road (solar panel).

To prevent emergence of hot spots in the solar panel, manufacturers install bypass diodes in the solar panels. When the weak cell hinders the flow of current, the bypass diodes get activated to provide it with an alternate route.

It is important to regularly clean the panels and ensure that they are free of shading and debris. Additionally, proper insulation and ventilation can help to dissipate heat and prevent overheating.

5.18 Bypass Diodes and Blocking Diodes

Bypass diode:

Bypass diodes are used in solar panels to prevent power loss caused by shading or malfunctioning of individual solar cells within the panel. When a part of the solar panel is shaded or a cell is damaged, the overall output of the panel decreases, and the shaded or damaged cells can act as load (resistors), causing energy dissipation as heat. This heat can further damage the cells and decrease the panel's efficiency.

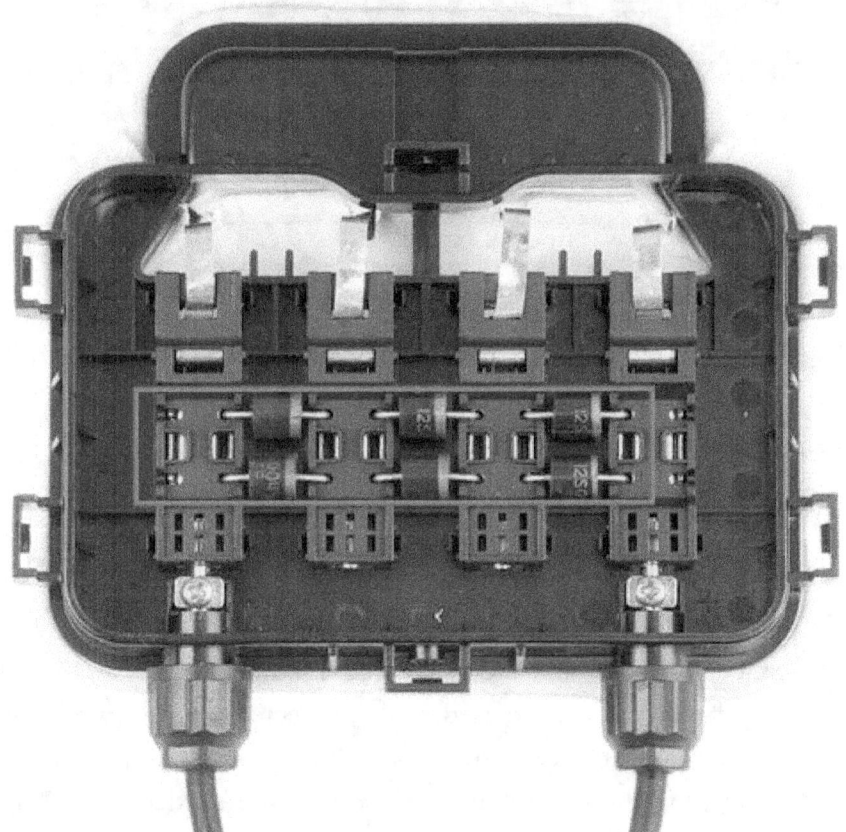

Fig 5.30

Source: GlobalSpec

A bypass diode is connected in parallel with a group of solar cells or a single solar cell within the panel. When a shaded or damaged cell causes a drop in voltage, the bypass diode provides an alternative current path, allowing the current to flow around the affected cell or group of cells, hence "bypassing" them. This prevents power loss, reduces heat dissipation, and improves the overall efficiency of the solar panel.

Example:

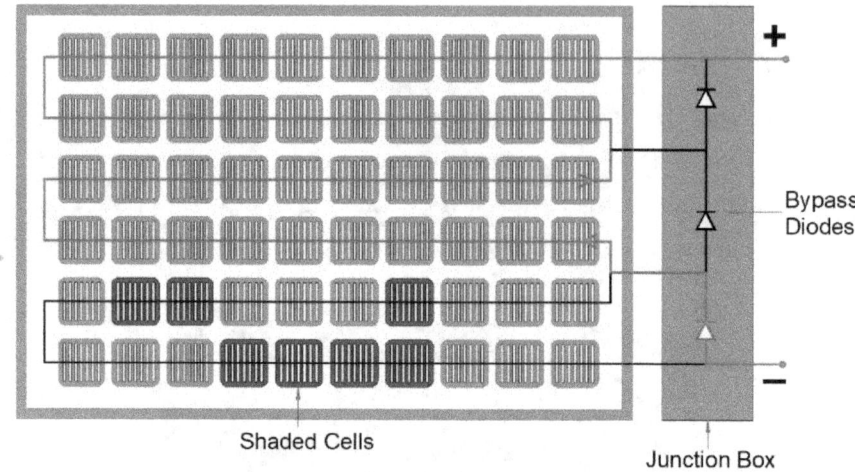

Fig 5.31

Imagine a solar panel with 60 solar cells connected in series. Under normal conditions, each cell generates power, and the overall power output is the sum of the power generated by all the cells. Now, suppose a tree's shadow falls on 6 of the 60 cells, causing a significant drop in the output voltage of those shaded cells.

Since the cells are connected in series, the shaded cells act as resistors, reducing the overall current and power output of the panel. Moreover, the shaded cells start dissipating heat, which can damage them and reduce the panel's efficiency.

To solve this issue, bypass diodes are connected in parallel with groups of solar cells. In our example, 3 bypass diodes are used, each connected to a group of 20 cells. When the tree's shadow falls on the 6 cells, the bypass diode associated with that group of 20 cells activates, allowing the current to flow around the shaded cells. This prevents power loss, reduces heat dissipation, and maintains the overall efficiency of the solar panel.

Blocking diode:

Blocking diodes are used in solar power systems to prevent the reverse flow of current from the battery back into the solar panel during the night or when the solar panel is not generating power. When a solar panel is not generating power, it can act as a load, and the battery can discharge through the panel, causing power loss.

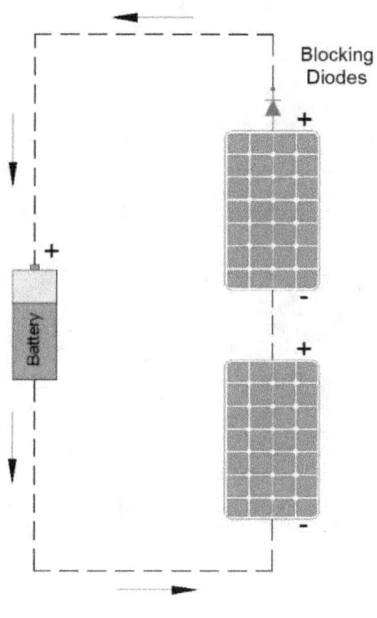

Fig 5.32

A blocking diode is connected in series with the solar panel. It allows current to flow only in one direction, from the solar panel to the battery. When the solar panel is not generating power, the blocking diode prevents the reverse flow of current, protecting the battery from discharging and ensuring the stored energy is used efficiently.

In the above setup, during the day the solar panel (at high potential) produces electricity and charges the battery (at low potential). During night, when the panel is not producing any electricity (low potential), the battery is at a higher potential. There is a possibility of the current flowing from the battery to the solar panel, thereby discharging the battery overnight. To prevent this from happening, a blocking diode is installed. It allows the current to flow from the panel to the battery but blocks the flow in opposite direction. It is always installed in series with the solar panel.

Chapter 6
Solar Charge Controller

6.1 What is Role of a Charge Controller?

A solar charge controller is an essential component of any solar power system that uses batteries for energy storage. Acting as the heart of your solar setup, it ensures the efficient and safe transfer of power from the solar panels to the batteries. In this chapter, we will explore the importance of solar charge controllers, discuss the two main types - Pulse Width Modulation (PWM) and Maximum Power Point Tracking (MPPT) charge controllers, and provide insights into choosing the right one for your system.

The primary function of a solar charge controller is to manage the process of charging batteries with the energy generated by solar panels. It performs several crucial tasks, including:

- **Regulating voltage and current**: The charge controller ensures that the batteries receive the appropriate voltage and current for optimal charging, preventing overcharging and maximizing battery life.

- **Protecting the batteries:** By preventing overcharging, deep discharging, and reverse current flow, the charge controller safeguards the batteries from damage and extends their lifespan.

- **Monitoring and controlling system performance:** Many modern charge controllers provide system monitoring features, allowing you to track the performance of your solar power system and make adjustments as needed.

Charge controllers play a crucial role in off-grid solar systems, as they manage the energy flow between solar panels and batteries. They are responsible for protecting batteries from overcharging, ensuring optimal charging, and extending battery life.

6.2 Types of Charge Controller

There are two main types of solar charge controllers: Pulse Width Modulation (PWM) and Maximum Power Point Tracking (MPPT) controllers. Each has its unique features, benefits, and limitations.

1. Pulse Width Modulation (PWM) Charge Controllers:

PWM charge controllers regulate the voltage from solar panels to match the battery bank's voltage. The controller works by regulating the flow of energy from the solar panels to the batteries using a technique called Pulse Width Modulation (PWM). In simple terms, the charge controller switches on and off rapidly, controlling the amount of energy delivered to the batteries by adjusting the duration of the "on" and "off" periods. This helps to maintain a stable charging voltage and keeps the batteries charged at an optimal level.

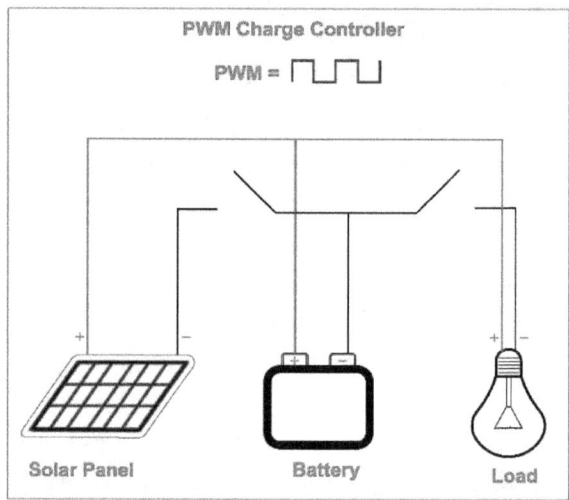

Fig 6.1

Some key features and benefits of PWM charge controllers are:

- **Cost-effective:** PWM charge controllers are generally more affordable compared to other types of charge controllers, such as Maximum Power Point Tracking (MPPT) controllers. This makes them a popular choice for smaller solar power systems or budget-conscious users.

- **Simplicity:** PWM charge controllers have a straightforward design and are relatively easy to use, making them suitable for beginners or users with less technical expertise.

- **Compatibility:** These charge controllers are compatible with a wide range of battery types, including lead-acid, gel, and AGM batteries.

- **Efficiency:** While not as efficient as MPPT charge controllers, PWM controllers still provide a decent level of efficiency in managing the charging process, ensuring that the batteries are charged safely and effectively.

It is important to note that PWM charge controllers are best suited for solar power systems where the solar panel voltage is close to the battery voltage. In systems with a significant voltage difference between the panels and batteries, an MPPT charge controller may be a better choice, as they offer higher efficiency and better performance under varying conditions.

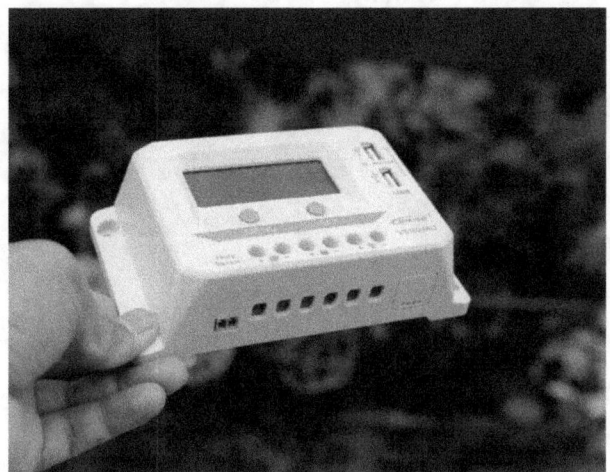

Fig 6.2

Here are some popular brands and models of PWM charge controllers:

1. Morningstar
2. Renogy
3. Blue Sky Solar
4. EPEVER
5. Schneider Electric

2. Maximum Power Point Tracking (MPPT) Charge Controllers

Maximum Power Point Tracking (MPPT) charge controllers are advanced solar charge controllers used in solar power systems to manage the process of charging batteries. Their primary function is to optimize the energy harvested from solar panels and ensure that the batteries are charged efficiently and safely. They are particularly beneficial in systems with higher power output and varying environmental conditions.

MPPT charge controllers work by continuously tracking the Maximum Power Point (MPP) of the solar panels, which is the point at which the panels produce the maximum amount of power. The MPP can vary due to changes in sunlight intensity, temperature, and other factors. By adjusting their operation to match the MPP, MPPT controllers can extract the maximum available power from the solar panels and convert it to the appropriate voltage for charging the batteries.

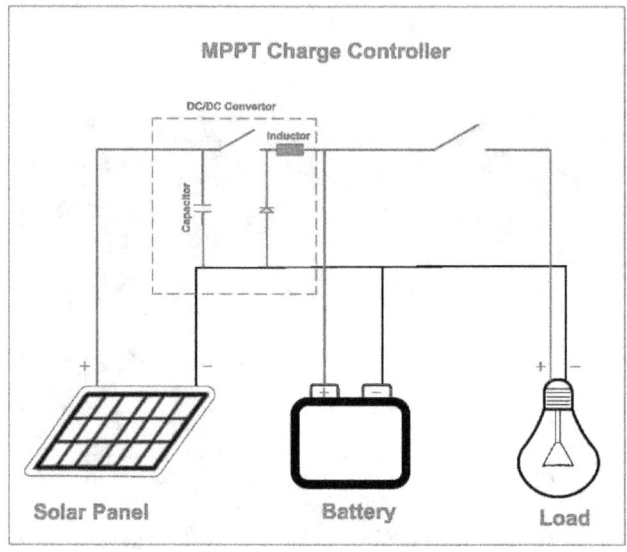

Fig 6.3

Some key features and benefits of MPPT charge controllers are:

- **High efficiency:** MPPT charge controllers are more efficient than PWM controllers, as they can extract more power from the solar panels, especially in variable weather conditions. This can result in up to 15-30% more energy harvested compared to PWM controllers.

- **Versatility:** MPPT controllers can handle a wider range of input voltages from solar panels, making them suitable for various solar power system configurations. They can also work with different battery types, such as lithium-ion, lead-acid, gel, and AGM batteries.

- **Improved performance in low-light conditions:** MPPT charge controllers are known for their ability to perform well in low-light or cloudy conditions, where other charge controllers might struggle to extract maximum power from the solar panels.

- **Scalability:** MPPT controllers are a better choice for larger solar power systems or systems that may be expanded in the future, as they can handle higher power output and offer greater flexibility in terms of system design.

The main drawback of MPPT charge controllers is their higher cost compared to PWM controllers. However, the increased efficiency and improved performance often justify the higher price for users looking to maximize their solar power system's output.

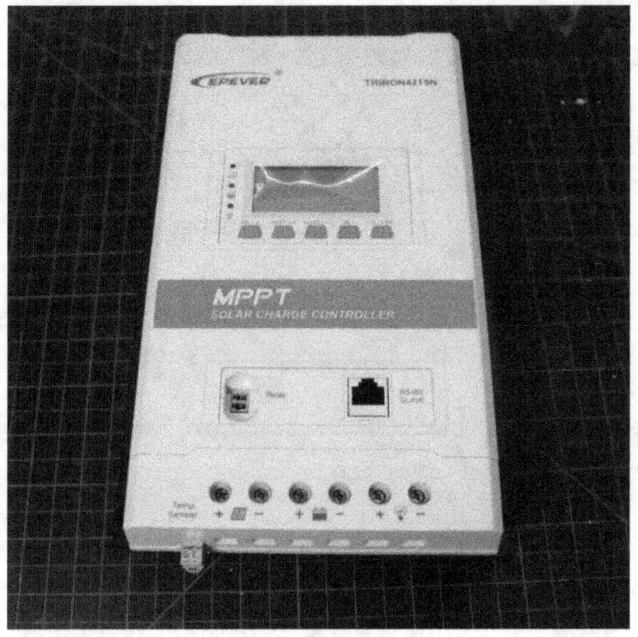

Fig 6.4

Here are some popular brands of MPPT charge controllers:

1. Victron Energy
2. MidNite Solar
3. OutBack Power
4. EPEVER
5. Schneider Electric

6.3 Charging Stages of Solar Charge Control

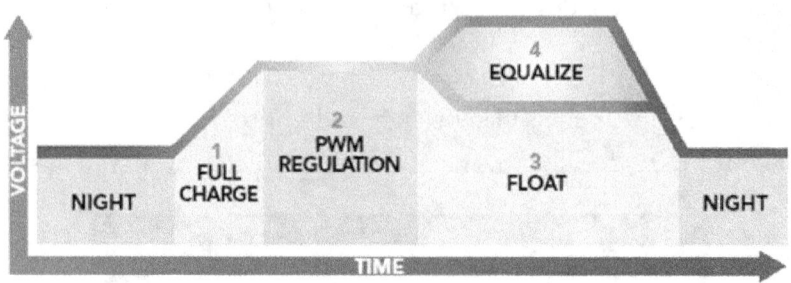

Fig 6.5

During the charging process, the charge controller measures the state of charge (SOC) of the battery. Based on the measured value, the charge controller increases or decreases the electric current to comply with particular stages of charge in the battery. The principal stages involved in the charging process are:

- **Bulk Charge:** In this stage, the solar charge controller allows the maximum current generated by the solar panels to flow into the batteries. The controller continuously monitors the battery voltage, and once it reaches a predetermined level, the charging process moves to the next stage.

- **Absorption Charge:** The absorption stage ensures that the batteries are charged to their maximum capacity. The controller maintains a constant voltage and gradually reduces the charging current to prevent overcharging. This stage ensures that the battery reaches a 100% state of charge (SOC).

- **Float Charge:** Once the battery is fully charged, the solar charge controller enters the float charge stage. In this stage, the controller maintains a lower voltage level to keep the battery at full charge, without causing overcharging or excessive gassing. This stage also compensates for self-discharge and keeps the battery ready for use.

- **Equalization Charge (optional):** Some solar charge controllers include an equalization stage, which is a controlled overcharge process that balances the individual cells in the battery. This stage is essential for flooded lead-acid batteries to prevent sulfation and prolong battery life.

6.4 Charge Controller Protection Features

A solar charge controller offers several protection features to ensure the safety and longevity of solar batteries and other components in the system. These include:

- **Overcharge Protection:** The controller prevents batteries from overcharging by regulating the charging current and voltage, ensuring that the batteries are charged safely and efficiently.

- **Deep Discharge Protection:** The controller disconnects the load from the batteries if the voltage drops below a preset level, preventing damage to the batteries due to deep discharge.

- **Overload and Short Circuit Protection:** The controller protects the system from overloads and short circuits, ensuring the safety of connected equipment and preventing damage to the components.

- **Reverse Polarity Protection:** This feature protects the controller and connected components from damage due to incorrect wiring or reverse polarity connection.

- **Temperature Compensation:** Some controllers have built-in temperature sensors that adjust the charging parameters based on the ambient temperature, ensuring optimal battery charging under varying temperature conditions.

6.5 Choosing the Right Solar Charge Controller

When selecting a solar charge controller for your solar energy system, it is essential to consider various factors to ensure that the controller meets the system's requirements and operates efficiently. By carefully examining the datasheet of the solar charge controller, you can determine if the controller is suitable for your solar energy system. The following steps can guide you in selecting the right charge controller using its datasheet:

- **Determine the system voltage:** Identify the voltage of your solar energy system. The most common system voltages are 12V, 24V, and 48V. The solar charge controller you choose should be compatible with the system voltage.

- **Calculate the solar array current:** To determine the appropriate current rating for your charge controller, calculate the total current generated by your solar panels. Add up the short circuit current (Isc) of each panel and multiply the sum by 1.25

to account for safety and environmental factors. This will give you the maximum current your solar array can generate.

- **Choose a charge controller with a suitable current rating:** Select a solar charge controller with a current rating equal to or greater than the maximum solar array current calculated in step 2. This ensures that the controller can handle the maximum current generated by your solar panels without getting damaged or overheating.

- **Decide between PWM and MPPT technology:** Solar charge controllers use either Pulse Width Modulation (PWM) or Maximum Power Point Tracking (MPPT) technology. MPPT controllers are generally more efficient and better suited for larger solar arrays or systems with varying solar conditions. However, PWM controllers are usually more affordable and can be suitable for smaller solar energy systems

- **Evaluate protection features:** Ensure that the solar charge controller has essential protection features, such as overcharge, deep discharge, overload, short circuit, and reverse polarity protection. These features are crucial for safeguarding your solar energy system components.

- **Consider additional features:** Look for additional features that may be beneficial for your solar energy system, such as temperature compensation, equalization charge, LCD display, remote monitoring and control, and compatibility with different battery types.

6.6 Example Calculation

Let us assume you have a 24V solar system with four solar panels, each rated at 200W

Fig 6.6

Step 1: Determine the system voltage

In this example, the system voltage (Battery Bank Voltage) is 24V.

Step 2: Determine the total solar panel wattage

Add up the power of each solar panel

Total Wattage = 4 x 200W = 800W

Step-3: Calculate the total current

Divide the total solar panel wattage by the system voltage

Current = Total Wattage / System Voltage

Current = 800/24 = 33.33A

Step 4: Add a safety margin:

It is a good practice to add a safety margin to the total current output to account for any potential fluctuations or inefficiencies in the system. A 25% safety margin is commonly recommended.

Total Current x 1.25 = 41.66A

In this case, the maximum solar array current is 41.66A.

Step 5: Choose a charge controller with a suitable current rating

Select a solar charge controller with a current rating equal to or greater than 41.66A. In this case, a 50A or 60A rated charge controller would be suitable.

For this example, you can select iTracer-ND Series (60A) MPPT Charge Controller from EPEVER.

Fig 6.7

Specifications:

System Nominal Voltage - 12 /24/36/48VDC auto

Battery Type -Lead-acid

Charging Technology - MPPT

Maximum PV Charging Current -60A

Maximum Discharge Current - 60A

Maximum PV Open Circuit Voltage -150V

Display - LCD and LED

Communication -RS485

Chapter 7
Off Grid Solar Inverter

The cornerstone of any solar energy system is the solar inverter, a vital component that plays a pivotal role in converting and managing solar energy. This chapter will delve deeper into the intricacies of off-grid solar inverters, their operation, the types available in the market, and crucial factors to consider when choosing the best fit for your off-grid solar system.

7.1 What is Role of an Inverter?

In an off-grid solar system, the energy harnessed from the sun by photovoltaic (PV) panels is initially produced as direct current (DC). However, our everyday appliances such as refrigerators, televisions, lights, and computers, operate on alternating current (AC). Herein lies the key role of a solar inverter - it is essentially a power adapter for solar energy, converting the DC power produced by your solar panels and stored in your battery bank into AC power that can be used by your appliances.

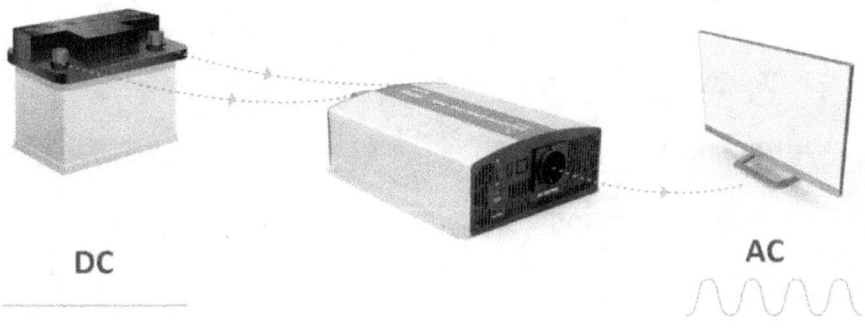

DC

AC

Fig 7.1

Source: Epever

In addition to this critical conversion process, a solar inverter manages the flow of electricity, intelligently distributing power between the solar panels, battery bank, and your home's electrical loads. It also safeguards your system against potential issues like overloading and overheating, thereby ensuring the smooth and safe operation of your off-grid solar system.

7.2 How Does An Off Grid Inverter Work?

An off grid solar inverter is a battery-based system. It is used for battery charging and load sharing. When the Sun is unavailable, such as during night hours, an off grid solar inverter is used to feed load through a charged battery.

Its working mechanism can be summarized as follows:

Case I - During Daytime:

- Solar panels absorb sunlight and convert it into direct current (DC) electricity.

- This DC electricity from the solar panels goes to a charge controller (if one is installed in the system), which regulates the voltage and current coming from the panels. The charge controller prevents overcharging, which can severely damage the batteries.

- The regulated DC electricity charges the battery bank. Simultaneously, the DC power is sent to the off-grid inverter.

- The off-grid inverter then converts this DC power into alternating current (AC) power, which is used to supply electricity to the household appliances.

- Any excess power that is not being used by the appliances continues to charge the batteries until they are full.

Case II - During Nighttime or Sunless Periods:

- When there is no sunlight (or insufficient sunlight), the solar panels stop producing electricity.

- The off-grid inverter begins to draw the stored DC power from the battery bank.

- The inverter converts this DC power into AC power to supply electricity to the household appliances.

- This process continues until the sun rises and the solar panels begin to produce electricity again, or until the batteries are depleted.

The role of the inverter in both these cases is vital to ensure a steady power supply for your off-grid living.

7.3 Types of Off-grid Solar Inverter

Off-grid solar inverters come in three primary types: Pure Sine Wave, Modified Sine Wave, and Square Wave inverters. Each varies in their output waveform, efficiency, compatibility with appliances, and cost.

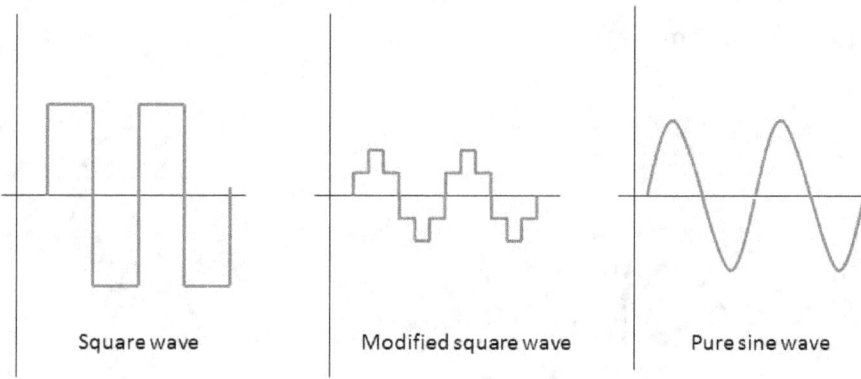

| Square wave | Modified square wave | Pure sine wave |

Fig 7.2

I.Pure Sine Wave Inverters:

These inverters produce power in a smooth, sinusoidal wave identical to the standard grid power, resulting in a quiet, hum-free operation of appliances. This type is the most expensive, but they are the most efficient and compatible with almost all electrical appliances, even the most sensitive ones.

II.Modified Sine Wave Inverters:

These produce a step-like waveform that approximates a pure sine wave. They are cheaper than pure sine wave inverters but are less efficient and could cause some devices like sensitive electronics, certain chargers, and some types of equipment to operate with less efficiency or even malfunction.

III.Square Wave Inverters:

These are the most basic and least expensive type, producing a square-shaped waveform. Due to their low efficiency and potential for high harmonic distortion, they can potentially damage sensitive electronics and are seldom used in modern off-grid solar systems.

7.4 Inverter / Charger

The term "Inverter/Charger" or "Combined Inverter Charger" refers to a device used in solar energy systems that integrates the functions of a solar charge controller and an inverter into a single unit. This unit is capable of both converting DC to AC and AC to DC. The combination of these functions is typically more cost-effective, efficient, and easier to install, commission, and maintain.

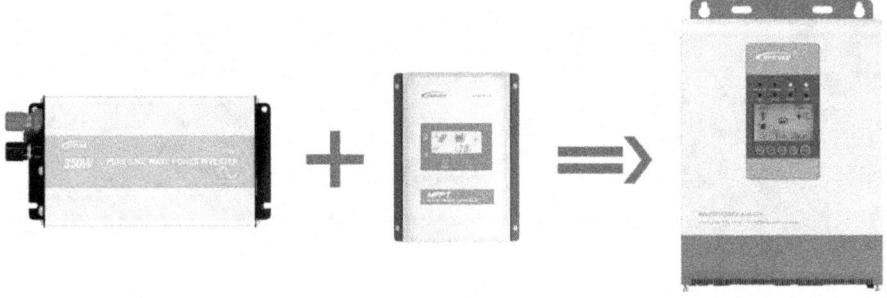

Fig 7.3

Source: Epever

In situations where there is limited solar radiation or unreliable grid power, an Inverter/Charger is an excellent option because it automatically switches between available sources such as solar panels, external power generators, and battery-stored energy to provide uninterrupted power output for consumers.

Furthermore, a combined charger inverter provides not only smart switching but also intelligent modes to optimize energy consumption and generation cycles and determine the energy budget balance.

Fig 7.4

Source: Epever

7.5 Inverter Power Rating

Inverter power rating, often referred to as the inverter size or capacity, is an important specification to consider when designing a solar power system. The power rating of an inverter is a measure of the maximum amount of electrical power that the inverter can supply to the load (appliances or equipment) at any given time. It is typically measured in watts (W) or kilowatts (kW).

Inverters come in a wide range of power ratings, from small units designed for RVs and boats that may be rated at just a few hundred watts, up to large commercial units rated in the tens of kilowatts or even higher.

When selecting an inverter, there are two power ratings you need to consider:

Continuous Power Rating:

This is the maximum amount of power that the inverter can supply on a continuous basis. This should be higher than the total power rating of all the devices you plan to run simultaneously.

A 1500W inverter can power up to 1500 Watts continuously. It is also called the nominal AC output power of the inverter.

Fig 7.5

Peak or Surge Power Rating:

This is the maximum amount of power that the inverter can supply over a short period, typically a few seconds to a few minutes. This power is required when appliances with electric motors (like refrigerators, air conditioners, or pumps) start up because they draw more power initially. The surge power rating of the inverter should be higher than the starting power requirement of your highest-draw appliance.

A 1500W inverter rated at 3000W surge watts can handle up to 4000 watts momentarily while starting things like motors.

7.6 Inverter Efficiency

Inverter efficiency is a critical factor to consider when selecting an inverter for your solar power system. It refers to how well an inverter can convert direct current (DC) from your solar panels or battery bank into alternating current (AC) for use in your home or business. The higher the efficiency, the more DC power is converted into usable AC power.

Inverter efficiency is typically given as a percentage. For instance, if an inverter has an efficiency rating of 95%, this means that 95% of the DC power input to the inverter is converted into AC power, while the remaining 5% is lost as heat.

It is important to note that inverter efficiency is not constant and can vary depending on several factors:

- **Load Level:** Inverter efficiency tends to vary with the amount of load on the inverter. Many inverters achieve their peak efficiency at around 75-80% of their maximum load. When the inverter is operating at lower loads (for instance, when fewer appliances are being used), its efficiency may decrease.

- **Power Factor:** The type of load (resistive, inductive, or capacitive) can also affect the efficiency of the inverter. Some loads, like motors and some electronics, can introduce phase shifts between current and voltage, leading to a lower power factor and reduced efficiency.

- **Temperature:** Higher operating temperatures can reduce the efficiency of the inverter and shorten its lifespan. This is why inverters are often equipped with cooling fans and should be installed in well-ventilated locations.

- **DC Input Voltage:** The efficiency of the inverter can also depend on the DC input voltage. It's important to match the input voltage from your solar panels or battery bank to the input voltage range of your inverter to ensure maximum efficiency.

7.7 Key Features of Off-Grid Solar Inverters

Off-grid solar inverters have a wide range of features which are mentioned below:

- **Overload and short-circuit protection:** They offer protection from damage due to short circuits and excess load, thus ensuring the longevity of the system.

- **Remote monitoring:** This feature allows us to monitor the performance of the inverter remotely, using a computer or a mobile device. We can keep an eye on the performance of our system, detect any issues or malfunctions, and take action before they become major problems

- **Automatic voltage regulation:** This feature ensures that the AC output voltage is stable and doesn't fluctuate, even if the load on the inverter changes. Thus ensuring that devices and appliances powered by the system receive a consistent and stable supply of electricity, which is crucial for their proper functioning.

- **Low battery protection:** This feature allows the inverter to shut down automatically when the battery voltage drops below a certain level, preventing it from getting damaged. This feature ensures that the battery remains healthy and extends its life, while also preventing any damage to the inverter itself.

Chapter 8
Essential Tools for DIY Solar Installation

8.1 Screwdrivers

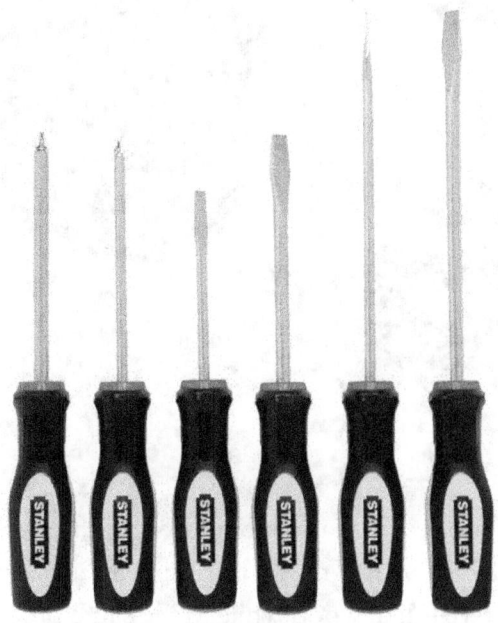

Fig 8.1

When working on off-grid solar applications, screwdrivers are vital tools for securing components, making electrical connections, and adjusting mounting hardware. Magnetic-tipped screwdrivers can be particularly useful for handling small screws and reaching tight spaces, preventing screws from falling or getting lost.

When working with electrical components, insulated screwdrivers are crucial for safety, as they feature non-conductive handles that minimize the risk of electrical shocks. For this purpose, purchasing a screwdriver tested to resist up to 1,000 Volts AC or 1,500 Volts DC is the best choice.

Fig 8.2

This is one tool which will never go unused and you don't necessarily have to be a big time Do-It-Yourselfer to get one of these. Impact drills are very useful to drive screws through heavy solid materials like wood while at the same time hammering them down into the metal or wood with a focused powerful torque.

Having a battery-powered drill or driver would allow you to complete electrical or construction work professionally while saving a ton of time. This comes as a handy tool when installing roof racks and mounts for solar panels. Mostly, roofs are at a slanted angle, making solar panel installation a safety hazard for installers. As compared to corded drills, battery-operated drills are efficient and most importantly portable.

8.3 Wire Strippers

Fig 8.3

Wire strippers is must-have tool for your off-grid solar installation. They are used to remove insulation from wires, exposing the conductor. They have multiple notches or cutting edges to accommodate different wire gauges, allowing you to strip wires without damaging them. These tools make it easy to prepare wires for connections and ensure a reliable electrical setup.

8.4 Cable Strippers

Fig 8.4

A cable stripper is designed to strip the outer insulation or sheathing from larger cables, typically used in electrical wiring. These cables may have multiple conductors within a single sheath. Cable strippers have adjustable blades that can accommodate various cable diameters and insulation thicknesses, allowing for precise stripping of the outer layers. It can cut PVC, rubber, foamed polyethylene (PE), and other insulating materials.

Cable strippers are suitable for larger cables with size in between from #5AWG to 4/0AWG, while wire strippers are designed for smaller wires with gauges between 10-24AWG.

The choice between a cable stripper and a wire stripper depends on the size and type of cables you are working with in your off-grid solar installation.

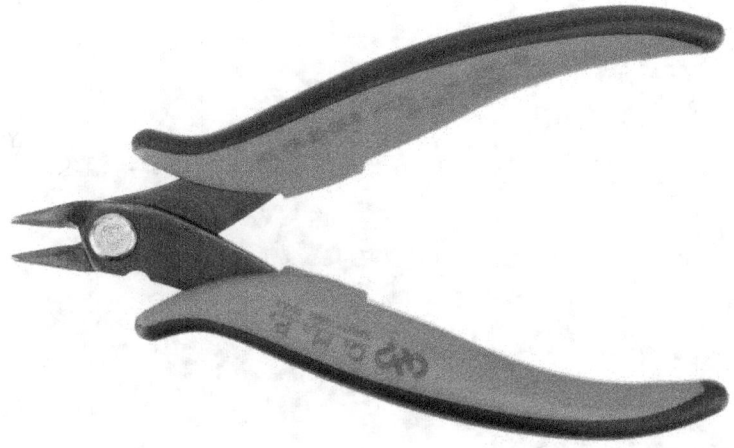

Fig 8.5

A wire cutter is a handy tool for cutting electrical wires during off-grid solar installations. You can use it to trim wires to the desired length, remove damaged sections, or separate wires during installation or maintenance. There are different types of wire cutters available, so choose one that suits your needs. Make sure it can handle the wire gauge you are working with.

Fig 8.6

Generally, wire cutter is best suited for cutting smaller size wires, but there may be where the wire cutter alone might not be enough to do the job for heavy-duty

119

applications with thicker cables. The cable cutter is the perfect choice for this purpose since it can cut up to 0AWG gauge cables (50mm²), both copper and aluminium.

8.6 Needle Nose Plier

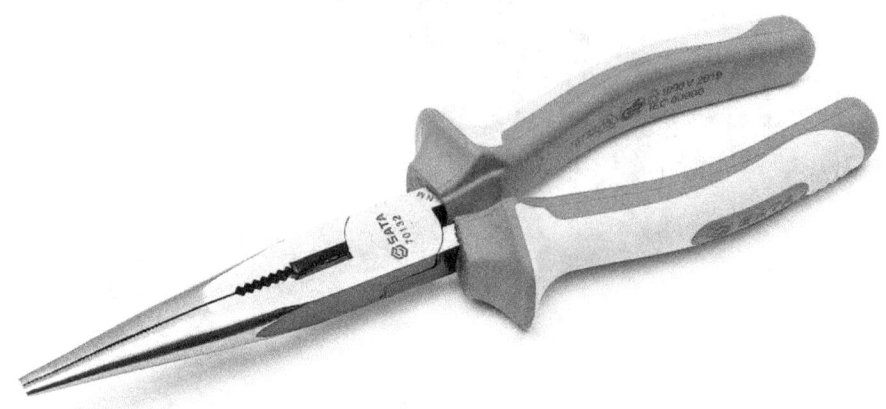

Fig 8.7

A needle nose plier is a tool used to bend and cut wires. It has long, thin ends that can reach into small spaces. It is really good when you have to work with small items or in tight spots. It also has a sharp part for cutting wires. But, it's not usually used for cutting; mostly for bending and handling wires. There are different types of handles for comfort and safety.

Always prefer to go with comfort-grip rubber handles that can withstand up to1000V and meets IEC standards.

8.7 Crimping tool

A crimping tool is an essential tool for making secure and reliable connections between wires and connectors. Crimping tools are specifically designed to compress and deform metal or plastic sleeves (crimp connectors) onto the ends of wires. This creates a tight and permanent connection that ensures proper electrical conductivity.

There are different types of crimping tools available, depending on the specific application and connector type. Common types include hand-held manual crimping tools and ratchet-style crimping tools, each with their own advantages and features.

i) Manual Crimping Tool:

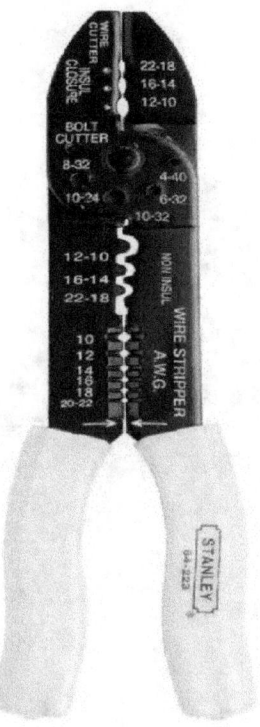

Fig 8.8

It comes in various types, including crimpers with fixed and interchangeable dies or jaws. They offer versatility by accommodating a wide range of connectors and wire sizes commonly used in off-grid solar application.

Manual wire crimping tools are used for crimping wire terminals with sizes in between 22 and 10AWG (0.5-6mm²). They are usually split into three crimping options marked by red, blue, and yellow, indicating the gauge ranges for each purpose.

ii) Ratchet-style Crimping Tool:

Fig 8.9

Ratchet-style crimping tools are commonly used for Solar panel and battery cable connectors. These tools provide consistent and reliable crimps by applying consistent pressure and ensuring full crimp completion before releasing. The ratcheting mechanism helps achieve a secure and uniform connection every time.

iii) Heavy Duty Crimping Tool:

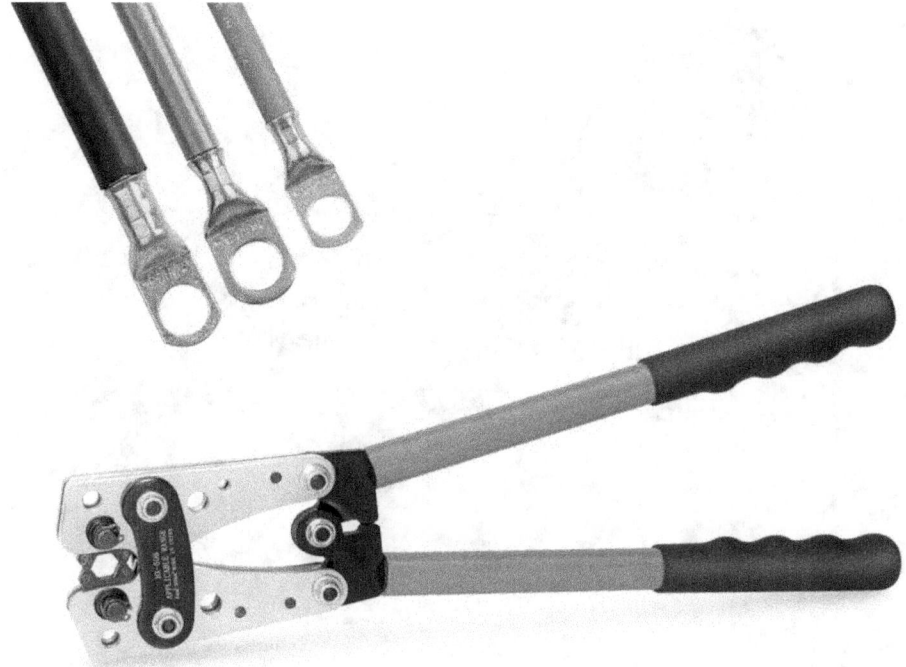

Fig 8.10

Battery cables are typically made of thick copper wire and have larger diameters compared to standard electrical wires. Therefore, you will need a heavy-duty crimping tool capable of crimping larger wire sizes, such as 2/0 or 4/0 gauge cables.

iv) Hydraulic Crimping Tools

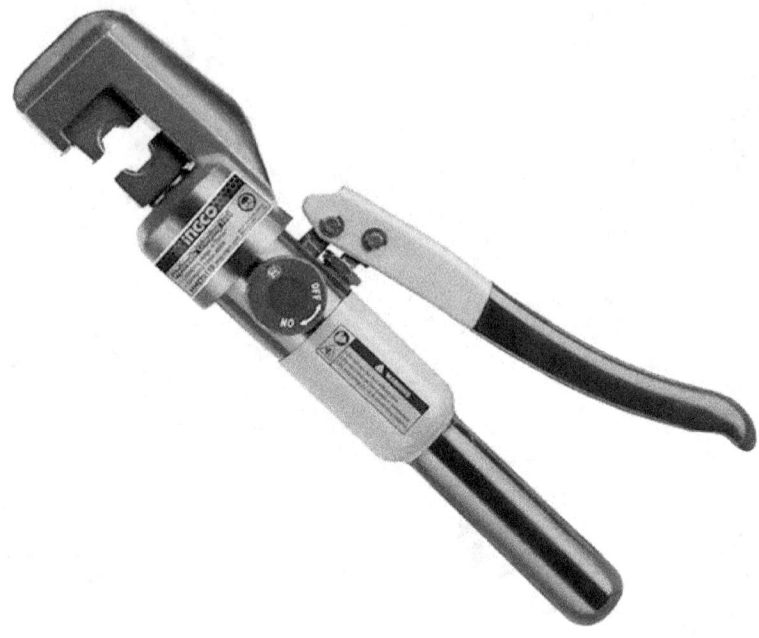

Fig 8.11

Hydraulic crimping tools provide compression power via a hydraulic fluid supply. They are designed to deliver maximum power and specially used for heavy-duty jobs. You may need this type crimping tools for thicker cable used in battery bank and battery bank to inverter.

8.8 Torpedo Level

Fig 8.12

A torpedo level is a nifty tool for making sure things are level and straight during off-grid solar installations. It is simple to use, you just place it on a surface or object, and the bubble inside the level will tell you if it is level or plumb. It is like a mini level with a couple of vials filled with liquid and an air bubble. You want to line up the bubble with the marked lines to know you are all set.

8.9 MC4 Spanners

Fig 8.13

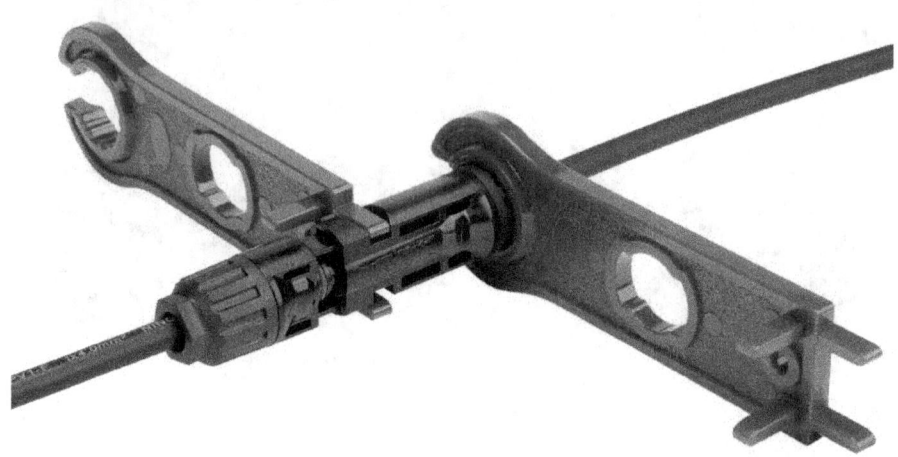

Fig 8.14

MC4 spanners are tools used to safely and easily connect or disconnect MC4 connectors. They are very useful for setting up or maintaining solar panels. These spanners help you get a firm grip on the connectors and prevent any damage when you are making or breaking the connection. They can also help you avoid electric shock by keeping your hands a safe distance from the connectors.

8.10 Hex Nut Ratchet Sets

Fig 8.15

Whether you are a professional or a DIY enthusiast, having a hex nut ratchet set in your toolkit is a smart investment. Hex nut ratchet sets are designed to accommodate a wide range of hex nut sizes, making them highly versatile. With a selection of interchangeable bits, these sets allow you to tackle different tasks without the need for multiple tools.

They are useful in making secure and reliable battery connections for the off grid solar system The ratcheting mechanism allows for easy tightening and loosening of battery terminals, saving time and effort.

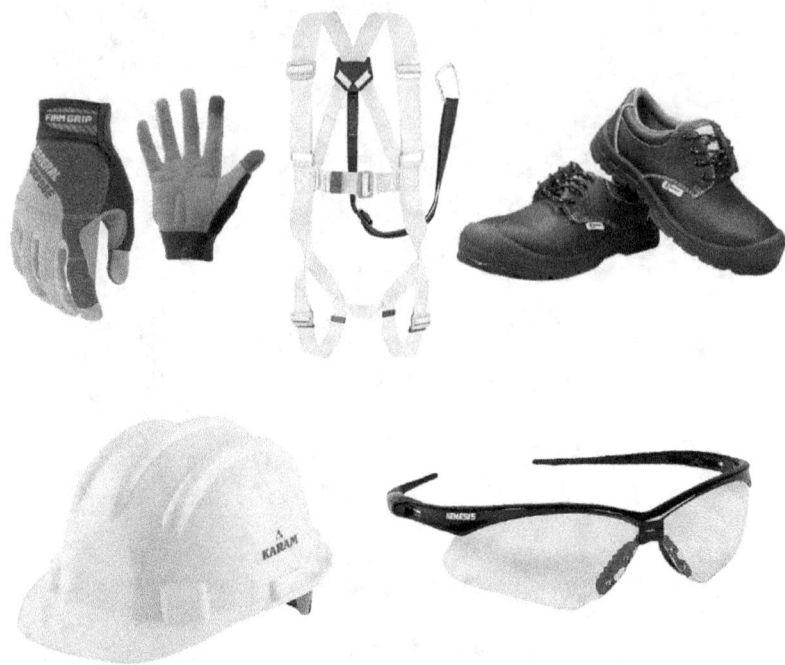

Fig 8.16

When it comes to installing a solar panel system, safety should be a top priority. Here are some casual tips on the essential safety equipment you will need:

1.   Safety harness:

If you are going up on rooftops, wearing a safety harness is a smart move. It'll help prevent any unwanted falls and keep you safe when you're working at heights of six feet or more.

2.   Safety glasses:

Do not forget to protect those peepers! Safety glasses shield your eyes from the glaring sun and any debris that could potentially cause accidents. Make sure they are ANSI Z87.1 approved and impact resistant for the best protection.

3.   Gloves:

Keep those hands safe and sound! Gloves not only give you a better grip but also protect your hands from cuts and scrapes. It is a no-brainer when it comes to safety.

4. **Safety helmet:**

When you are working at heights, it is always a good idea to protect your noggin. A safety helmet can come to the rescue if there is a fall, so keep it on for added safety.

5. **Protective footwear:**

Your feet need some love too! Opt for steel-toe shoes with slip-resistant outsoles when climbing ladders and manoeuvring on potentially slippery roof tiles. That way, you will have better traction and reduce the risk of slipping.

Remember, safety gear is crucial to ensure your well-being and the well-being of those around you during a solar panel installation. So gear up, stay safe, and get solar power flowing!

Chapter 9
Equipment Used in Off Grid Solar System

9.1 Solar Panel Connectors

Connectors are responsible for establishing secure electrical connections between various components of the system, such as solar panels, inverters, and charge controllers, ensuring seamless energy transfer and optimal performance.

An MC4 connector is used to connect solar panels with DC cables that carry DC power from the panels to the inverter. These connectors are the only certified and authorised options to do so. The process has to be so spotless that specialised MC4 crimping tools are needed to make this connection.

1. MC4 Connectors

MC4 (Multi-Contact 4 mm) connectors are used to connect solar panels with DC cables that carry DC power from the panels to the inverter. They are designed to withstand harsh outdoor environments and are resistant to water, dust, and UV radiation.

Some key features of MC4 connectors include:

- Rated Voltage: Up to 1,500 V DC
- Rated Current: Up to 50 A
- Operating Temperature Range: -40°C to 90°C
- Locking Mechanism: Snap-in design with a locking mechanism that requires a tool to disconnect
- Wire Size Compatibility: 10 to 12 AWG (American Wire Gauge)

Types of MC4 Connectors:

MC4 connectors are available in different types and configurations to suit various solar PV installations. The most common types of MC4 connectors include:

I.Male and Female MC4 Connectors:

Fig 9.1

MC4 connectors are gendered, with male and female counterparts that connect to form a secure connection. The male connector has a metal pin, while the female connector has a metal socket. These connectors are typically used to connect solar panels, inverters, charge controllers, and other system components.

II.MC4 In-Line Fuse Connectors:

Fig 9.2

In-line fuse connectors are designed to protect solar PV systems from overcurrent conditions by integrating a fuse within the MC4 connector. They are typically available in male and female configurations and various fuse ratings to match the specific requirements of the solar PV system.

III.MC4 T-Branch Connectors:

Fig 9.3

T-branch connectors are a combination of MC4 connectors and T-branch connectors, allowing for easy and efficient parallel connections between multiple solar panels. They are available as a pair of connectors, one male-to-dual female and one female-to-dual male, making it simple to connect two solar panels in parallel.

IV.MC4 Y-Branch Connectors:

Fig 9.4

Similar to T-branch connectors, Y-branch connectors combine MC4 connectors with Y-branch connectors to create multiple parallel connections simultaneously. They are available in various configurations, such as dual male-to-single female, dual female-to-single male, or multiple-input-to-single-output connectors.

V. MC4 Extension Cables:

Fig 9.5

These are pre-assembled cables with MC4 connectors at both ends, used to extend the distance between solar panels and other system components like inverters or charge controllers. They are available in different lengths and wire sizes to suit specific installation requirements.

2. Amphenol H4 Connectors

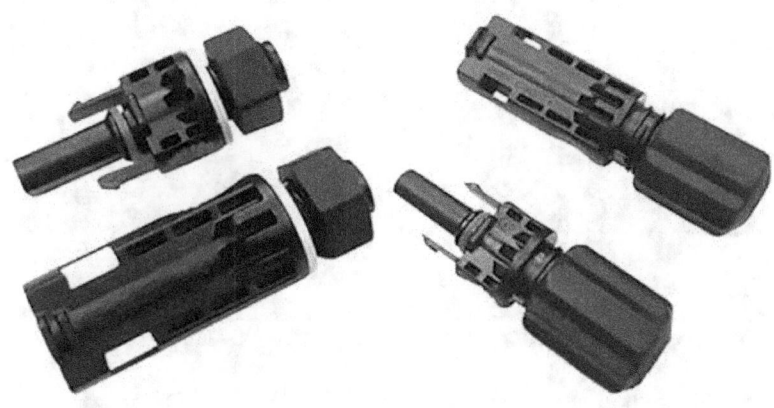

Fig 9.6

Amphenol H4 connectors are another popular option for solar projects, offering high current capacity and excellent environmental protection. They are often used in larger solar systems and are compatible with a wide range of solar panels. Some key features of Amphenol H4 connectors include:

- Rated Voltage: Up to 1,500 V DC
- Rated Current: Up to 65 A
- Operating Temperature Range: -40°C to 90°C
- Locking Mechanism: Snap-in design with a locking mechanism that requires a tool to disconnect
- Wire Size Compatibility: 10 to 12 AWG

9.2 Crimp Connectors

In a DIY off-grid solar system, you are working with a lot of electrical wiring. The way these wires connect to various components like your solar panels, charge controller, battery bank, and inverter largely determines the efficiency and safety of your system.

A weak or loose connection can result in energy loss due to resistance, overheating, and worst-case scenario, electrical fires. Crimping forms a secure metal-to-metal contact between the wire and terminal, ensuring optimal electrical conductivity and a solid connection.

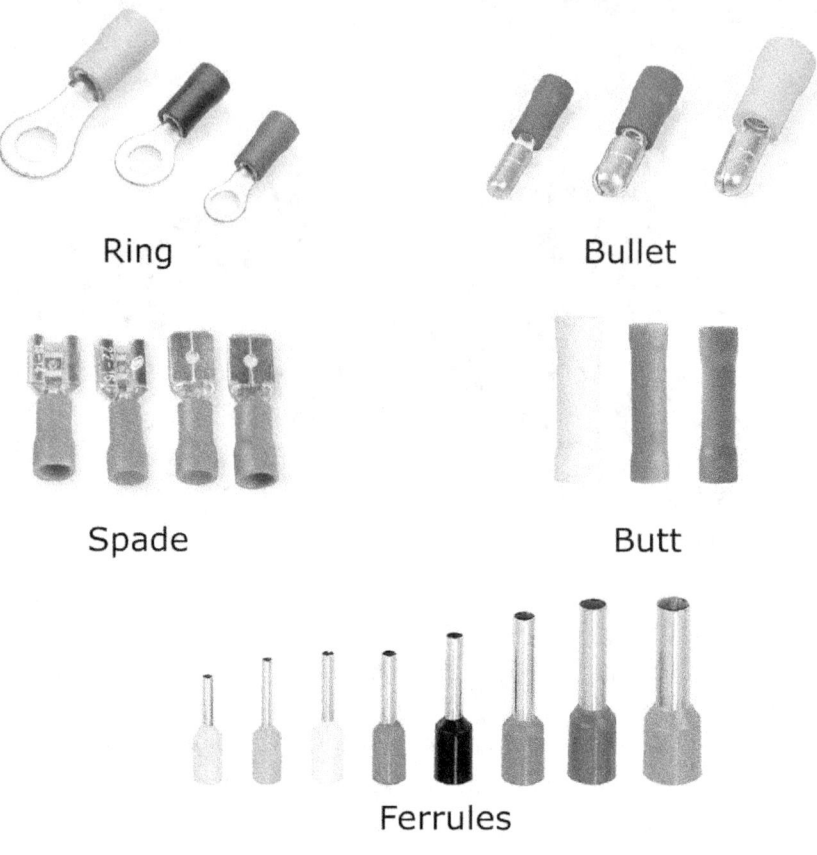

Ring

Bullet

Spade

Butt

Ferrules

Fig 9.7

They come in numerous shapes and sizes, depending on the specific use. Here are some of the most common types:

I. **Ring Terminals:**

These terminals have a circular end that's perfect for securing under a screw or bolt. They provide a solid connection and are commonly used in situations where the connection must not come loose easily.

II. **Spade Terminals:**

Also known as fork or split ring terminals, they are handy when you need to frequently disconnect and connect wires, as they can easily be slipped on and off a screw or stud.

III.Butt Connectors:

They are used to connect two wires end-to-end, making them great for wire splices or extensions. After inserting the wires, both ends of the butt connector are crimped to secure the connection.

IV.Bullet Terminals:

These terminals, named for their shape, are used for quick disconnect features where one wire has a male terminal (the bullet) and the other wire has a female terminal.

V.Ferrules:

These are used for terminating stranded wires, preventing the wire strands from fraying or breaking away and can be used to connect the wire to terminal blocks.

9.3 Cable Lugs

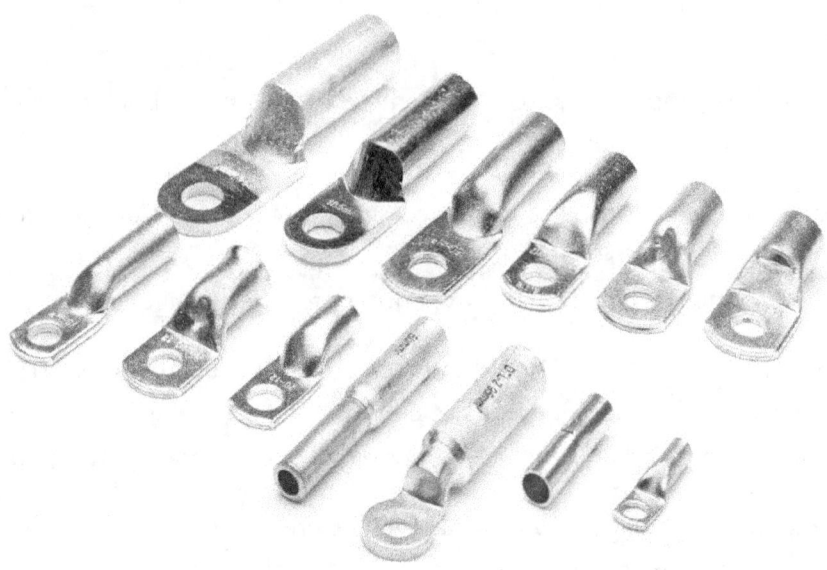

Fig 9.8

Image Source: electrical-cable

Cable lugs or terminal lugs, are connectors used to terminate the end of a cable or wire. Lugs are generally used for big cables only and for smaller wires, crimp connectors are used.

The primary purpose of a lug is to provide a secure, mechanical and electrical connection between the cable and the equipment it is connected to. They are commonly used to attach the cables coming from your solar panels to the charge controller and from there to the battery bank and the inverter.

Choosing the right lugs for your off-grid solar system is a key step towards a safe and efficient setup. To make a correct selection, you must consider the wire gauge and pick a lug that matches its size, as a mismatch could result in a loose or improper connection.

The lug's material, usually copper or aluminium needs to match that of the wire to avoid galvanic corrosion. Look for UL-listed wire lugs because cheap lugs might not have thick enough material.

Once you have crimped your lug onto the wire, a heat shrink tube is slipped over the lug and the adjoining section of the wire. When heat is applied, the tubing contracts tightly around the lug and wire, forming a seal that is resistant to water, dust, and other contaminants. This reduces the chance of breaking the individual wire strands in sharp bends.

9.4 Fuses

Fuses are used to protect your solar system from electrical abnormal situations like overcurrent and short circuits.

A fuse is a device that protects against excessive current flow in an electrical circuit. It contains a metal wire or strip that melts when too much current flows through it, thereby interrupting the circuit and preventing potential damage or fire.

Inside the fuse is a metal strip that is connected to both metal ends of the fuse body. If there is a short or fault anywhere in the circuit, or the circuit is overloaded, the metal strip, or link, heats up and quickly melts, opening the circuit and shutting off the power.

Fuses should be placed as close as possible to the energy source. If current flows from your battery to your inverter put the fuse as close to the battery as possible. If current flows from solar panels to the charge controller, place it close to the solar panels. Only place fuses on the positive (red) wire.

Fuses are typically used in the following areas in a solar system:

- Between the solar panels and the charge controller.
- Between the charge controller and the battery.
- Between the battery and the inverter.

Types of Fuses:

There are different types of fuses, but for solar applications, DC (direct current) rated fuses are used. Here are a few types of fuses frequently used in off-grid solar installations:

Cartridge Fuses:

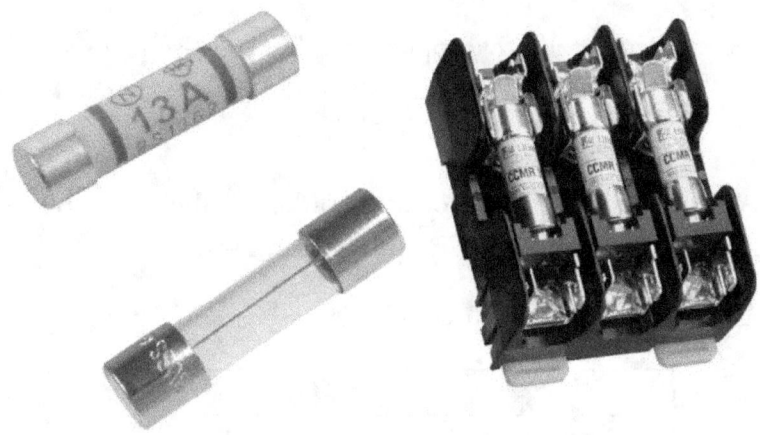

Fig 9.9

These are the typical glass or ceramic cylinder type of fuses with metal end caps. They are often used in low-current (< 30A) solar applications.

Blade Fuses:

Fig 9.10

These fuses are popular in automotive applications but can also be used in low-current solar applications. They are named for the flat, blade-like metal terminals protruding from the body.

Blade fuses are categorized and color-coded according to their amperage rating. This standardization makes it easier to identify and replace fuses.

Here is a general guideline for blade fuse ratings and their corresponding colour codes:

Current	Colour
2A	Grey
3A	Violet
4A	Pink
5A	Tan / Orange
7.5A	Brown
10A	Red
15A	Blue
20A	Yellow
25A	Clear / White
30A	Green
35A	Purple
40A	Amber / Orange

Remember, these colour codes apply to standard ATO, ATC, mini, low profile mini, and maxi blade fuses. The Micro2 and Micro3 blade fuses have a slightly different colour coding system.

NH Fuses:

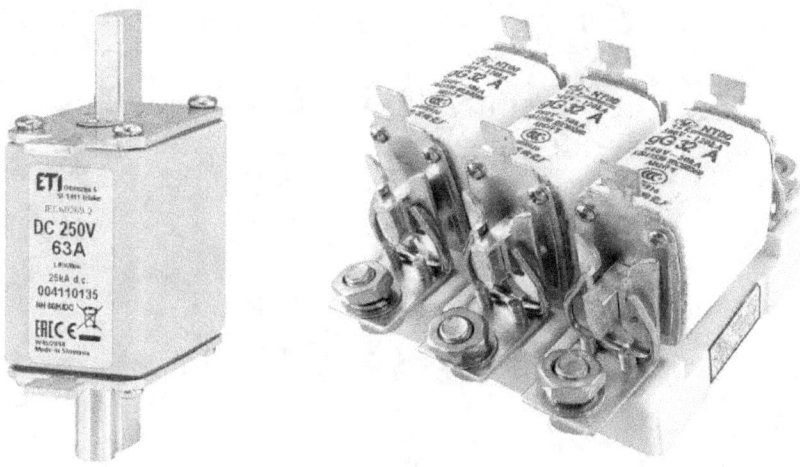

Fig 9.11

These fuses, also known as 'knife blade' or 'DIN' fuses, are commonly used in larger commercial and utility-scale solar installations. They are designed to interrupt high levels of current and can be mounted in disconnect and combiner boxes.

Midget Fuses:

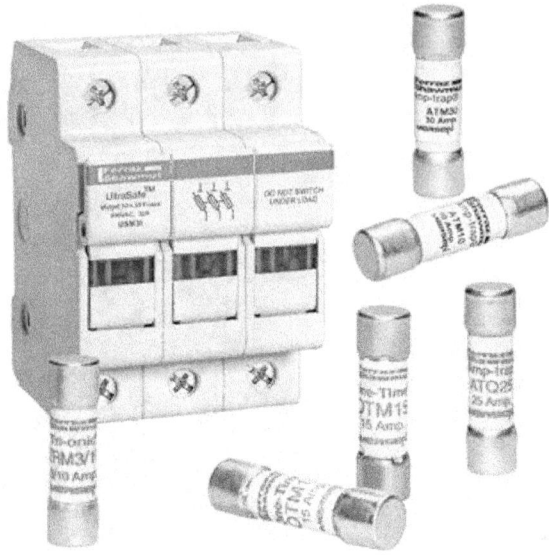

Fig 9.12

Despite the name, these fuses are designed for heavy-duty applications. They are cylindrical, similar to cartridge fuses, but have a larger diameter and are used in a variety of commercial and industrial solar installations.

Photovoltaic (PV) Fuses:

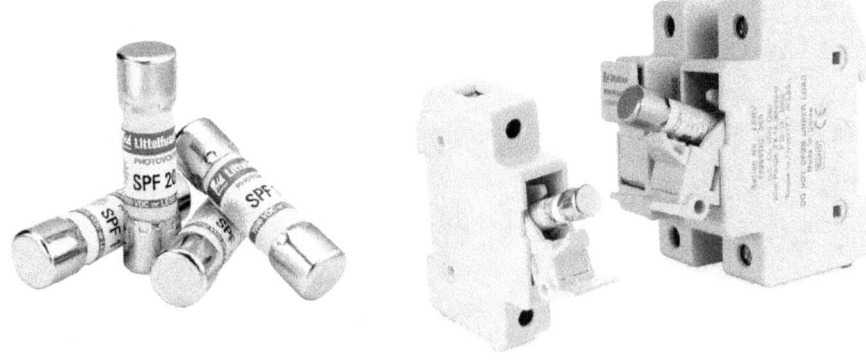

Fig 9.13

These are specialized fuses explicitly designed for solar PV applications. PV fuses are capable of interrupting the highest fault currents associated with solar panels and strings. They come in various form factors, including cartridge and midget, and are typically used in combiner boxes and inverters.

9.5 Circuit Breakers

A circuit breaker is a switch that automatically interrupts electrical flow in a circuit in case of an overcurrent condition. They are essential components of any electrical network as they prevent damage to equipment and fire outbreaks that might be caused by electrical faults. In an off-grid solar system, a circuit breaker is used to protect your solar panels, battery, and inverter.

Like fuses, circuit breakers also protect your system from overcurrent. The difference is that a circuit breaker is reusable. When it trips, you can reset it and use it again.

Circuit breakers are typically used in the same areas as fuses. They offer the advantage of acting as a disconnect, meaning you can manually open the circuit without needing to remove any components.

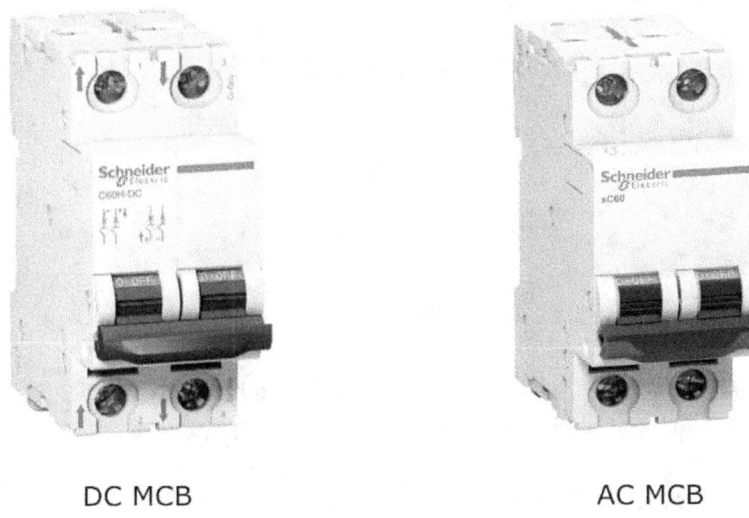

DC MCB   AC MCB

Fig 9.14

Here are the types of circuit breakers commonly used in off-grid solar power systems:

DC Circuit Breakers:

These are designed specifically for direct current (DC) applications. They are used in the DC side of a solar PV system, protecting the solar panels and batteries. In off-grid solar power systems, these are used between the solar panels and charge controller, as well as between the battery and inverter.

AC Circuit Breakers:

These are used in the alternating current (AC) side of a solar system to protect the inverter and AC loads. They are often used between the inverter and the AC distribution panel.

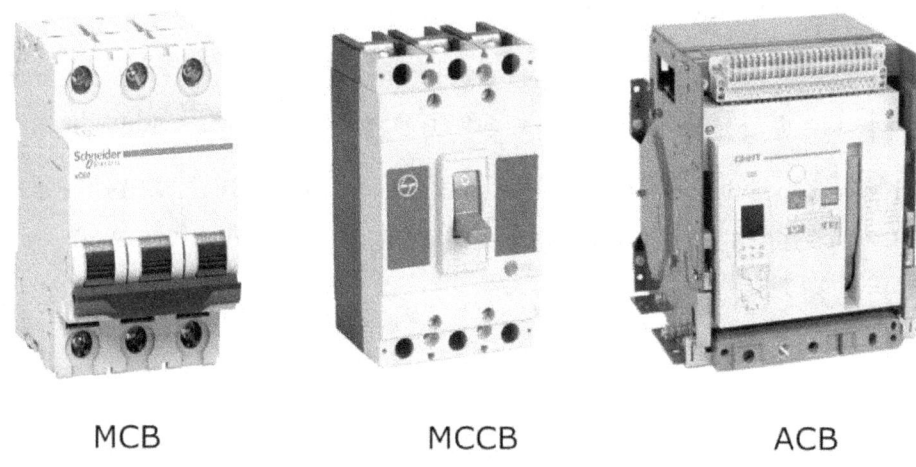

MCB   MCCB   ACB

Fig 9.15

Different types of DC and AC circuit breakers include:

- **Miniature Circuit Breaker (MCB):**

A commonly used circuit breaker that protects against short circuit and overload conditions. It is typically used for currents up to 100 Ampere.

- **Molded Case Circuit Breaker (MCCB):**

Used for higher current ratings, typically up to 1000 Ampere. They can also be adjusted to trip at varying current levels.

- Air Circuit Breaker (ACB):

Generally used for high current applications (over 1000 Ampere) and offer the highest interrupting capacity.

146

Circuit breakers can also be classified based on the number of poles or the number of circuit conductors they can protect. This classification is very important as it indicates the number of circuits a breaker can handle, or in the case of a fault, the number of circuits it can interrupt.

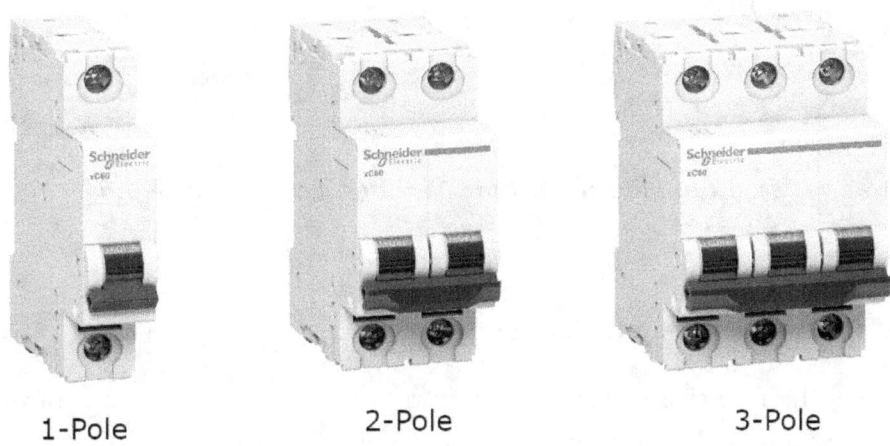

1-Pole   2-Pole   3-Pole

Fig 9.16

- **Single, Double, or Triple Pole Breakers:**

Single-pole breakers protect one circuit, double-pole breakers protect two circuits or a single two-phase circuit, and triple pole breakers protect three circuits and are commonly used in three-phase systems.

Always Use DC Fuses/Circuit Breaker in DC solar System

Solar systems typically generate direct current (DC) electricity, so the fuses you use must be rated for DC power. The challenge with DC is that, unlike alternating current (AC), which passes through zero voltage points during each cycle where the current can naturally be interrupted, DC power maintains a constant voltage. This makes it more challenging to interrupt the current flow in a DC system because there is no natural 'break' in the power.

DC-rated fuses and circuit breakers are specifically designed to cope with these challenges. They are constructed in a way that more effectively extinguishes the electrical arc, providing a secure means of interrupting DC current.

So, always make sure you are using DC-rated fuses in your DC solar system. Using an AC-rated fuse in a DC system can be dangerous.

Sizing Of Fuses / Circuit Breakers

It is essential to select a fuse / circuit breaker rated for a voltage higher than the maximum system voltage. The current rating of the fuse should be 125% to 156% of the expected maximum current. The detailed is given below.

DC Breaker / Fuses:

As per NEC, the DC fuse or breaker size can be determined as per the following equation:

$$Circuit\ ampacity = Short\ Circuit\ Current\ (Isc)\ X\ 1.56$$

Example-1:

A 315 Watt module with an Isc rating of 9.12A.

To calculate the fuse size required between the string and the charge controller, you take 9.12 x 1.56 = 14.7 and round up to the next trade size of 15A.

Example-2:

Fuse between the battery bank and inverter = (continuous Watts / Battery Voltage) x 1.56

A 1000W /12V inverter draws = 1000/12 =83.3A,

Circuit Ampacity = 83.3 x 1.56 = 130A, round up to the next standard trade size which will be 150A.

AC Breaker/Fuses:

AC Breaker is placed at the inverter output and the Outlet for AC appliance.

The above NEC ampacity formula also changes on the AC side of the circuit. Instead of 1.56, the multiplier is 1.25. And in place of the short-circuit current, you must use the maximum or continuous output current listed on the inverter specification sheet.

$$Circuit\ Ampacity = Inverter\ AC\ output\ current\ X\ 1.25$$

Example-3:

let's assume a 1500W inverter with an AC output of 6.5A max.

Circuit Ampacity = 6.5 x 1.25 = 8.12A, round up to the next standard trade size which will be 10A.

9.6 DC Isolator Switch

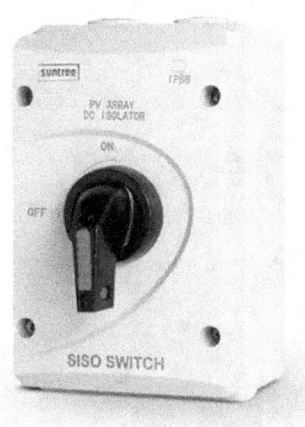 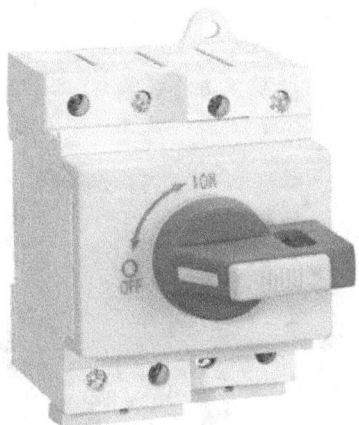

Fig 9.17

A DC Isolator Switch is a device that allows for the safe interruption of the direct current (DC) produced by the solar panels. This type of switch is crucial for maintenance, troubleshooting, or emergency situations as it provides a means to safely disconnect the electrical power.

In a typical solar PV setup, you will find a DC isolator switch on the roof next to the solar panels and another one near the inverter. The rooftop isolator allows for safe maintenance of the panels themselves, while the one near the inverter provides a way to cut the power before it reaches the inverter, protecting it during maintenance or in case of a fault.

Before you buy a DC isolator switch, make sure it complies with the system's current and voltage. The isolator switch used between the solar panels and the charge controller will typically handle a lower current, as solar panels often output a relatively low current. However, as the voltage from the solar panels can be quite high, especially in series configurations, the isolator switch must be rated for these higher voltages.

On the other hand, the isolator switch used for the battery connection will need to handle higher currents, as batteries can discharge at high current levels, but usually at lower voltages.

It is important to note that in many regions, the use of DC Isolator Switches in solar PV installations is mandated by electrical safety regulations. Therefore, they are a critical component in the design and operation of solar PV systems.

9.7 DC Bus bars

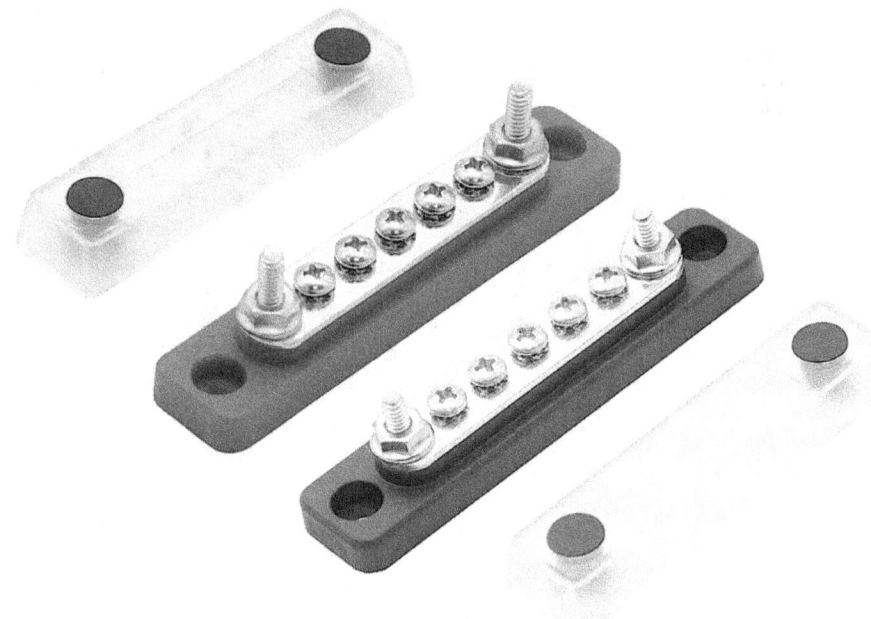

Fig 9.18

Bus bars are used to distribute Electrical Power among various loads. It takes the power from the source. After that, it transfers the power to the other loads connected to it, bus bars are the point where multiple loads or multiple sources are connected.

It is usually a thick strip or bar of copper, brass, or aluminium, known for their low resistance, that serves as a common connection point for multiple circuits.

In an off-grid solar power system, typically three types of busbars are used to facilitate safe and efficient power distribution: positive busbars, negative busbars, and earth (or ground) busbars. Each type serves a distinct purpose:

- **Positive Busbars:** These are used to collect and distribute positive (+) DC voltage coming from solar panels or batteries. They connect all positive terminals in a system and lead them to various devices such as charge controllers or inverters.

- **Negative Busbars:** These are used to consolidate and route the negative (-) DC voltage. They link all the negative terminals and offer a return path for the electrical current after it has passed through the loads (like lights or appliances).

- **Earth/Ground Busbars:** They connect the metal cases of your devices to the earth, ensuring that in case of a fault (like a wire touching the case), the electrical current goes safely into the ground instead of causing harm.

9.8 Surge Protection Device (SPD)

Fig 9.19

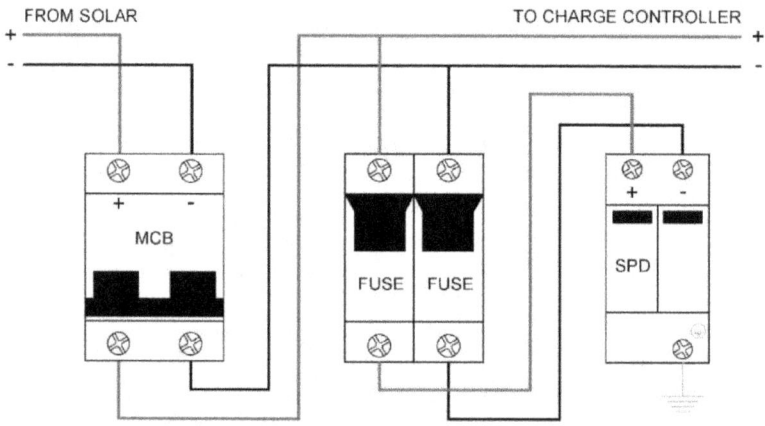

Fig 9.20

Surge Protection Device (SPD) is used to protect the system's electrical equipment from potential damage caused by electrical surges or spikes, which can be triggered by factors such as lightning strikes or power surges on the grid.

The solar SPD is designed to divert the excess voltage from a surge safely to the ground, thus protecting the rest of the system. It is a critical component for the longevity and safety of a solar PV system, protecting expensive components like inverters and charge controllers from being damaged by high voltage spikes.

There are typically two types of SPDs used in a solar PV system:

- **Type 1 SPD:**

Used in situations where direct lightning strikes are possible. It is installed in the main distribution board and protects against very high-energy surges.

- **Type 2 SPD:**

Used for protection against indirect lightning strikes or surges from the grid. It is installed at the inverter input and output, as well as the charge controller input.

9.9 Combiner Box

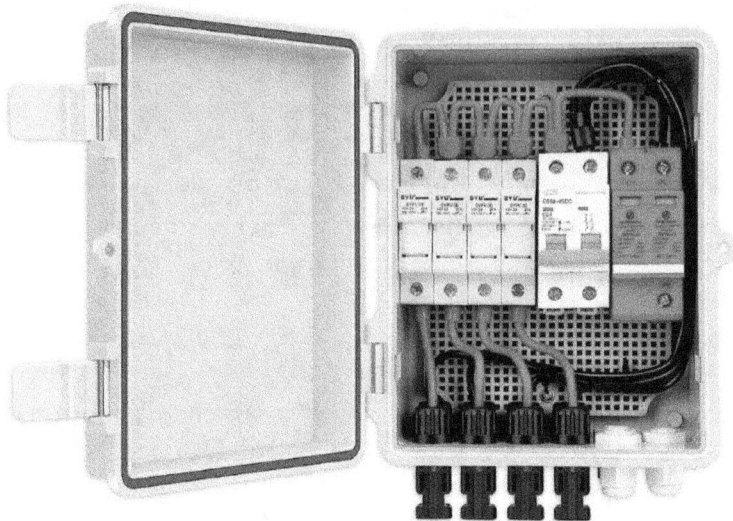

Fig 9.21

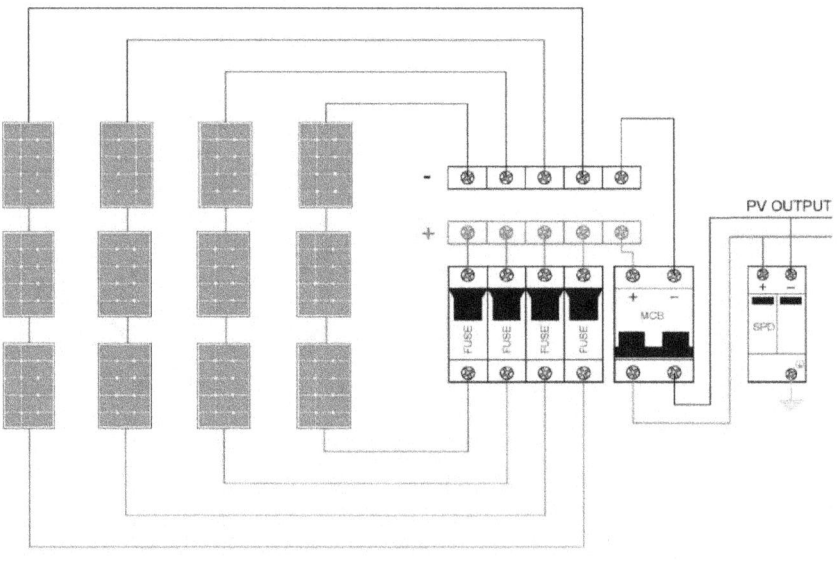

STRING COMBINER BOX WIRING DIAGRAM

Fig 9.22

A Combiner Box is an important component of a solar photovoltaic system, especially in larger systems that have multiple solar panel strings.

The Combiner Box is located between the solar panels and the inverter or charge controller. Its primary function is to bring together the output of several solar strings into a single output. This reduces the number of wires that need to be run and simplifies the wiring system overall.

Key features and functions of a Combiner Box include:

Combining Outputs:

The main function of a combiner box is to combine the output of several solar strings into one. This is particularly important in larger installations where there may be dozens of individual solar panels, each producing a relatively small amount of power.

Safety and Protection:

Each string input in the combiner box has its own circuit breaker or fuse. This is a safety measure to prevent overcurrent conditions that can occur if there is a fault or issue with one of the solar strings. The breakers or fuses can cut off the problem string without affecting the operation of the rest of the system.

Monitoring and Diagnostics:

Many combiner boxes also include monitoring systems. These can provide information about the performance of each solar string, making it easier to identify and address any problems.

Chapter 10
Earthing and Lightning Protection

Solar installations, like all electrical systems, require comprehensive safety measures to protect both the equipment and the people using it. Among these safety measures, earthing (grounding) and lightning protection hold significant roles. This chapter aims to elucidate these essential topics, offering practical insights to guide you through the hows and whys of implementing effective earthing and lightning protection systems.

10.1 What is Earthing (Grounding)?

Earthing, or grounding, is the practice of connecting electrical systems directly to the earth using conductive materials. This serves to ensure that in the event of a fault or surge, excess electrical energy has a safe path to flow into the earth, minimizing the risk of electrical shock or fire.

10.2 Why is Earthing Important?

Imagine your off-grid solar setup as a mini power station. You've got all these cables, devices, and panels generating and transmitting electricity. With all these components, things can easily go sideways, and that's where grounding comes in as your unsung hero. So, let's break down why you should care:

1. Personnel Safety

One of the primary reasons for earthing is to protect people from electrical shocks. An ungrounded system can become a hazardous voltage source if a live wire comes into contact with a non-electrical component, like the metallic frame of a solar panel or an appliance casing. By grounding these elements, any stray electrical energy is safely channeled into the earth, reducing the risk of electric shock to those interacting with the system.

2. Equipment Protection

Electrical surges or spikes can occur for various reasons, such as lightning strikes or malfunctions in electrical components. These surges can damage sensitive electronics and other appliances connected to the system. Grounding provides an alternative path for excessive electrical energy, diverting it safely into the earth and thereby protecting your equipment.

3.   Voltage Stabilization

Grounding ensures that all parts of the electrical circuit are at the same voltage level, relative to the earth. This uniform voltage is essential for the proper operation of electrical and electronic devices, contributing to their longevity and reliability. It also aids in preventing voltage fluctuations that could otherwise cause intermittent operation or failure of devices.

4. Fault Clearance

A well-grounded system supports the proper operation of overcurrent devices like fuses and circuit breakers. When a fault occurs, grounding helps to ensure that enough current flows to "trip" these protective devices, cutting off the electrical supply and thereby preventing further damage or injury.

5. Electromagnetic Interference

Grounding can reduce the electromagnetic interference (EMI) in your system. EMI can disrupt the proper functioning of sensitive electronics and data lines. A proper grounding scheme can act as a shield, offering some level of protection against EMI.

6. Legal Requirements

Finally, proper grounding is often not just a best practice but a legal requirement as well. Electrical codes in many jurisdictions mandate grounding for safety reasons, and failure to comply can result in fines, legal repercussions, and voided insurance policies.

10.3 Two Types of Grounding

There are two types of grounding: chassis (or mechanical) and electrical. Both are completely different grounds and must be understood correctly. 'Ground' is just a reference point; it does not always mean neutral.

Setting up correct chassis and electrical grounds is essential for safety, efficiency, and the longevity of systems. Chassis grounds connect all exposed metal parts that don't carry current (like the solar module frame, battery case, back plate, and mounting structures) together and to the ground. This is mainly done for safety. It ensures that if there's a build up of electrical potential on any metal part, it won't give you a shock when touched. Keep in mind, voltage is essentially the difference in potential between two points. Properly grounding your system negates this difference, ensuring safety

10.4 Key Components to Ground in an Off-Grid Solar System

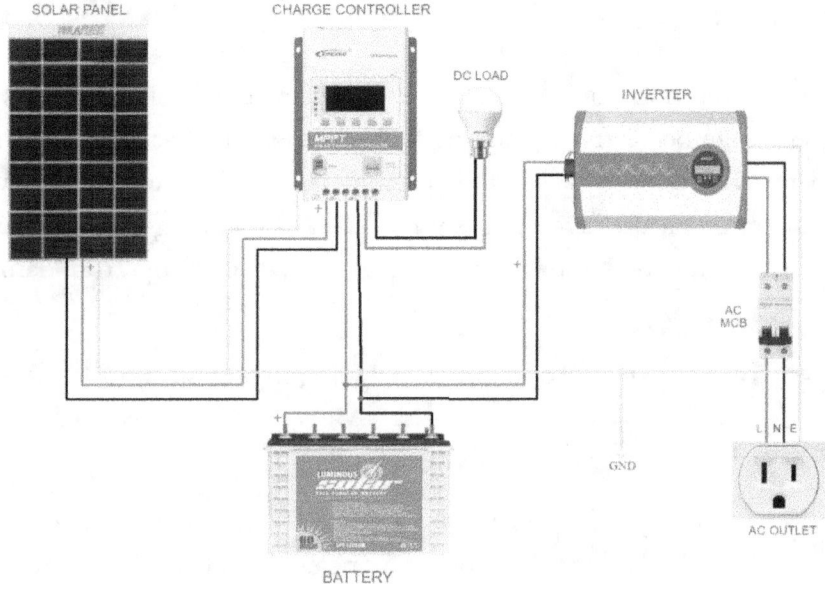

Fig 10.1

1. Solar Panels: The metal frames of all solar panels should be grounded to ensure that they do not carry a dangerous voltage if a fault occurs.

2. Inverter: Inverters should be grounded to ensure safety and proper functioning.

3. Racking/Mounting Equipment: The metal parts of your racking system should also be grounded.

4. Battery Bank: Particularly if you're using metal-encased batteries.

10.5 Typical Grounding Method for Mobile Setups

For off-grid mobile setups, such as vans or boats, it's essential to link all metallic components to the primary grounding point. This includes the boat's exterior hull or a van's main metal structure. When establishing this link in vehicles, prioritize connecting to the robust structural frame over the flimsier side panels. Prepare the connection point by sanding away any paint, then bore a hole to attach a wire to the central grounding bar. Subsequently, fasten another wire from this bar to the battery bank's main negative terminal.

The ground wire should match the diameter of its corresponding active wire. However, manufacturers may have varying specifications on this topic. It's advisable to consult the user manuals of both the charge controller and the inverter to ensure correct wire sizing.

10.6 Lightning Protection

Lightning is a common cause of failures in solar photovoltaic (PV) systems. A damaging surge can occur from lightning that strikes a long distance from the system, or even between clouds. But most lightning damage is preventable. A lightning protection system typically involves a combination of lightning rods, conductors, and ground electrodes designed to protect a structure and its contents from damage due to electrical surges.

Below are the principal elements of a typical lightning protection system:

1. **Air Terminals (Lightning Rods)**

Air terminals, commonly referred to as lightning rods, are metallic devices installed at the highest points of a structure. Generally fabricated from conductive materials such as copper or aluminum, these rods function to attract and capture the electrical discharge from a lightning strike.

2. **Conductors**

Conductors are essentially the electrical pathways that facilitate the routing of electrical energy from the air terminals to the ground rods. These are often robust wires made from materials that offer low electrical resistance, typically copper or aluminum.

3. **Grounding Electrodes (Ground Rods)**

The grounding electrodes, or ground rods, are cylindrical rods embedded into the earth. Their primary function is to dissipate the electrical energy absorbed by the air terminals into the ground. These rods are usually constructed from highly conductive materials like copper or galvanized steel.

4. **Bonding**

Bonding is a critical aspect of a comprehensive lightning protection strategy. This involves the interconnection of all metallic objects in close proximity to the air terminals and conductors. The purpose of bonding is to minimize the risk of side flashes, which are secondary lightning strikes that occur when the electrical discharge seeks alternative pathways to the ground.

5. **Surge Protection Device (SPD)**

Fig 10.2

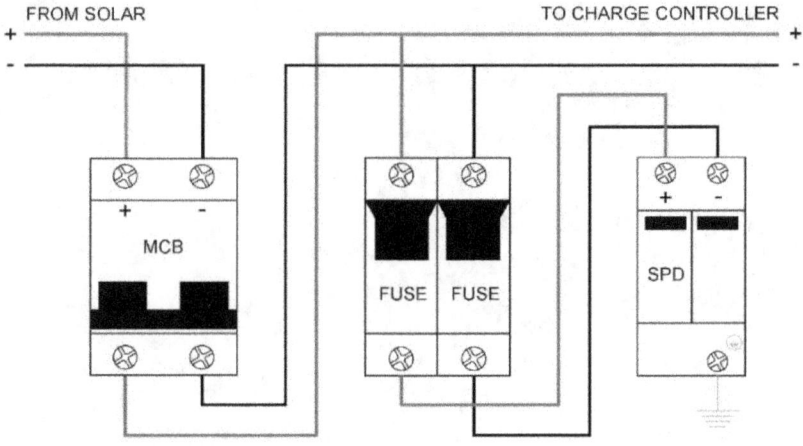

Fig 10.3

Surge Protection Device (SPD) is used to protect the system's electrical equipment from potential damage caused by electrical surges or spikes, which can be triggered by factors such as lightning strikes or power surges on the grid.

The solar SPD is designed to divert the excess voltage from a surge safely to the ground, thus protecting the rest of the system. It is a critical component for the longevity and safety of a solar PV system, protecting expensive components like inverters and charge controllers from being damaged by high voltage spikes.

10.7 DIY or Professional Installation?

Doing it yourself can be rewarding but consider professional installation for grounding and lightning protection. It is not just about hammering a rod into the ground; it involves calculating the correct rod length, appropriate wire gauge, and positioning of all elements for maximum safety.

Chapter 11
Sizing of Off-Grid Solar Components

Sizing an off-grid solar power system involves several steps to ensure that the system will be able to meet the electrical demand of your application, provide power during periods of low sunlight, and have a long enough lifespan to be cost-effective. Here is a step-by-step guide to sizing off-grid solar components:

11.1 Daily Energy Consumption

First, you need to calculate the total watt-hours per day which you will be using. This is done by listing all the electrical devices you plan to use, noting the power (in watts) each device requires, and multiplying that by the number of hours you plan to use them each day.

For example:

If you're running 2 Nos of 6W LED bulbs for 5 hours a day, 1 No of Fan (80W) for 4 hours, 1 No of Laptop (65W) for 3 hrs, and a WiFi Router (6W) for 24 hours.

1. LED Bulb: 2 x 6W x 5 hr = 60Wh

2. LED TV: 1 x 65W x 3 hr = 195Wh

3. Ceiling Fan: 1 x 80W x 4 hr = 320Wh

4. WiFi Router: 1 x 6W x 24 hr = 144Wh

Total = 719Wh = 0.719kWh /day

11.2 Solar Panel Sizing

Step 1: Determine Your Energy Needs

Let us consider the daily energy consumption is10kWh

Step 2: Calculate Peak Sun Hours

Peak sun hours refer to the number of hours during the day when the sun is strong enough to generate electricity. In more technical terms, it is the number of hours per day during which the average solar irradiance (sunlight) is 1000 watts per square meter (W/m2) or 1 kilowatt per square meter (kW/m2).

One peak sun hour = 1000 W/m2 or 1kWh/m2 of sunlight

For New Delhi, the approximate peak sun hour is 4.5 Hours

If you are situated in North America, you can use this reference chart, to get the number of peak sun hours.

Refer to my article on solar panel sizing to know more details on calculating Peak Sun Hours.

Step 3: Calculate the Total Watt-peak rating needed

Divide the daily energy consumption by the peak sun hours. This will give you the minimum number of watts your solar panel system needs to generate to meet your daily energy needs.

- Daily Energy Consumption is 10kWh x 1000 = 10000Wh
- Peak Sun Hours is 4.5

So the Watt-peak required = Daily energy consumption (Wh) / Peak Sun Hour

Panel Wattage Required = 10000/ 4.5 = 2500 Watt

Step 4: Consider Solar Panel Inefficiency Factors

The output of solar panels drops slightly each year, which is outlined by their performance warranty. If your solar panel's performance warranty guarantees 80% performance after 25 years, then their degradation rate is calculated as 20%/25 years, or 0.8% production loss each year. By the end of its lifecycle, a 500W-rated panel would only output 400 watts.

In addition, solar panels are tested in STC (Standard Test Conditions) i.e. cell temperature of 25°C and irradiance of 1000 W/m2 with an air mass of 1.5 (AM1.5)s. In the real world, solar panels often fall short of these lab-tested conditions, meaning they produce a bit less power than their wattage rating.

Because of the above factors, it is wise to include some extra solar capacity so that you can reach your target generation after accounting for the inefficiencies of the system.

20-30% is a good amount of headroom to account for inefficiencies.

Multiply your solar array size by 1.30 (130%) to account for this:

2500W x 1.3 = 3250 Watt

Step 5: Calculate the Number of Solar Panels Required

In the previous step, you have already calculated the Total Wattage of the Solar Array. Now you have to determine the number of solar panels that are required for your installation.

Currently, solar panels available in the market are in the range of 330 to 670 watts. (Smaller solar panels are not considered)

To determine the number of solar panels you need, divide the total Wattage of the Solar Array by the wattage of each panel.

In our example case, we need a system that needs to generate 3250 Watts per day and each panel selected is 540 watts

So, no of panels = Total Wattage Required / Rating of each Panel = 3250/ 540 = 6.01

You need six number of 540W panels.

11.3 Battery Bank Sizing

Batteries are sized based on the total energy storage required and the depth of discharge (DoD) that the batteries can safely handle.

Step 1: Determine the desired days of autonomy

Decide how many days of autonomy you want for your system, which is the number of days your battery bank should be able to supply power without being recharged by solar panels. This is particularly important for locations with inconsistent sunlight or during periods of bad weather. For example, let us assume you want 2 days of autonomy.

Step 2: Calculate the required battery capacity

Multiply your daily energy consumption by the desired days of autonomy to determine the required battery capacity:

Battery Capacity needed = Daily Energy Consumption x Days of Autonomy

From the example given in 11.1, the daily energy consumption is 719Wh.

Battery Capacity = 719 Wh/day x 2 days (autonomy) = 1438 Wh

Step 3: Account for battery depth of discharge

To prolong battery life and ensure optimal performance, it is important not to discharge batteries below a certain level. For lead-acid batteries, a typical DoD is about 50%, while for lithium-ion it can be up to 80-90%.

For LiFePO4 batteries with an 80% DoD:

Total Battery Capacity = 1438 Wh / 0.8 (80% discharge depth) = 1797.5 Wh

Step 4: Convert watt-hours to ampere-hours

To size the battery bank, you need to convert the required capacity from watt-hours to ampere-hours (Ah) by dividing the required capacity by the battery bank voltage. For example, if you choose a 12V battery bank:

1797.5 Wh / 12V = 149.79 Ah

In this example, you would need a 12V battery with a capacity of at least 150Ah to meet your daily energy needs and provide 2 days of autonomy.

11.4 Sizing of Charge Controller

Step 1: Determine the total solar panel wattage:

First, calculate the total wattage of your solar panel array by adding up the wattage of each panel. For example, if you have four 250-watt panels, the total wattage would be 1000 watts.

Step 2: Calculate the total solar panel current output:

To find the total current output, divide the total solar panel wattage by the system voltage. Most residential systems use either 12V, 24V, or 48V systems. For instance, if you have a 1000-watt solar array and a 24V system, the total current output would be 41.67 amps (1,000 watts ÷ 24 volts).

Step 3: Add a safety margin:

It is a good practice to add a safety margin to the total current output to account for any potential fluctuations or inefficiencies in the system. A 25% safety margin is commonly recommended. To calculate this, multiply the total current output by 1.25. In our example, the adjusted current output would be 52.09 amps (41.67 amps × 1.25).

Sample Calculation

Let us take a datasheet of a solar panel and calculate the required rating of the Charge Controller.

Consider a 100W solar panel used to charge a 12V battery bank.

1. Watt Rating:

The power rating of the panel is 100W

2. Current Rating:

Let the system voltage is 12V

So current = 100W / 12V = 8.33A

3. Consider Safety Margin:

Charge controller rating = Current Rating x Safety Factor = 8.33 x 1.25 = 10.41A

Therefore, the solar charge controller rating is selected next higher rating available in the market, which is either 15A/12V or 20A/12V.

11.5 Sizing and Selecting an Inverter

Let us consider a hypothetical off-grid solar system to illustrate the process of sizing an inverter. Assume the following devices and appliance will be powered simultaneously:

1. Refrigerator: 150 watts
2. LED lights (10): 10 watts each, totaling 100 watts
3. Laptop: 45 watts
4. TV: 100 watts
5. Microwave: 1,000 watts

Step 1: Calculate the total power consumption.

Add up the wattage of all the devices and appliances you plan to use simultaneously:

150 W (refrigerator) + 100 W (LED lights) + 45 W (laptop) + 100 W (TV) + 1,000 W (microwave) = 1395 watts

Step 2: Add a 20% safety margin

To account for fluctuations and surges in power demand, add a 20% safety margin to the total power consumption:

1395 W x 1.2 = 1674 watts

In this example, you would need an inverter with a power rating of at least 1674 watts.

Step 3: Consider surge capacity.

Check the surge or peak power of each device, which is often higher than the continuous power consumption. For example, the microwave might have a surge power of 1,500 watts. Ensure that the inverter's surge capacity can handle the combined surge power of all devices.

11.6 Sizing Fuses and Circuit Breakers

Here is a simple example of how to size a fuse or circuit breaker for a solar system.

Suppose you have a 12V solar system with a solar array producing a maximum of 40A current. You need to protect the system components from overcurrent and short circuits, so you must choose the right fuse or circuit breaker rating.

Step 1: Determine the maximum current

First, identify the maximum current flowing through the circuit that needs protection. In this case, it is 40A from the solar array.

Step 2: Apply an Overload Factor

Consider an Overload factor of 25% (1.25)

Step 3: Apply a safety factor

A safety factor ensures that the fuse or circuit breaker does not trip during normal operation. A common safety factor is 1.25 (125%).

Step 4: Calculate the Fuse Rating:

Fuse Rating = Current x Overload Factor x Safety Factor

$$= 40A \times 1.25 \times 1.25 = 63A$$

Step 5: Choose the nearest standard fuse or circuit breaker rating

Fuses and circuit breakers come in standard ratings, so you must select the nearest available size that is equal to or greater than the calculated minimum rating.

Chapter 12
Cable Selection and Sizing

12.1 Types of Cables Used in Off-Grid Solar Systems

Various types of cables are used in off-grid solar systems, each with specific applications and purposes. Understanding the different types of cables and their applications will help you choose the right cable for each component of your solar system. The cables used in these systems can be broadly categorized into two groups: DC cables and AC cables.

Fig 12.1

A. DC Cables

These cables handle the direct current (DC) generated by solar panels and are stored in batteries. They include:

- **PV Module Cables:**
  These cables connect the solar panels to the charge controller, which regulates the flow of power to the battery bank. PV module cables are typically 10-12

AWG (American Wire Gauge), double-insulated solar cables designed to handle the DC output from solar panels.

- **Battery Cables:**
  Battery cables connect the battery bank to the charge controller and the inverter. They are responsible for carrying the DC power between these components. Battery cables are generally larger in size, ranging from 2-4/0 AWG, depending on the system capacity and the current they need to carry.

- **Inverter Cables:**
  These cables connect the inverter to the battery bank, transferring the DC power from the batteries to the inverter. Inverter cables are usually similar in size to battery cables, typically 2-4/0 AWG, to handle the required current between the battery bank and the inverter.

- **Charge Controller Cables:**
  These cables connect the charge controller to the solar panels and battery bank. They should be sized according to the current and voltage requirements of the charge controller, as well as the distance between the components. Similar to battery cables, they are typically made of thick copper or aluminum wire with durable insulation.

B. AC Cables

These cables handle the alternating current (AC) produced by the inverter and distributed it to the electrical loads. They include:

- Inverter Output Cables:
  Inverter output cables transmit electricity from the inverter to the main electrical panel or distribution board. The appropriate AC wire size should be chosen in compliance with local electrical codes to ensure safety and efficiency.

- Distribution Cables:
  Distribution cables are responsible for distributing electricity from the main electrical panel or distribution board to various electrical loads or appliances within the system. They should follow local electrical codes and be appropriately sized based on the expected current and voltage requirements of the connected devices.

12.2 Factors to Consider When Selecting Cables

A. Cable size

Cable size is a crucial factor to consider when setting up an off-grid solar system, as it directly affects the system's efficiency, safety, and overall performance. Selecting the appropriate cable size involves taking into account the following aspects:

i. **Voltage drop:** Voltage drop refers to the reduction in voltage as electricity travels through a cable. To maintain efficient power transmission and minimize energy loss, it's important to limit the voltage drop. For DC cables in solar systems, aim for a voltage drop of less than 3%, while for AC cables, a drop of less than 5% is acceptable.

ii. **Current carrying capacity:** The cable size should be chosen based on its ability to carry the maximum current expected in the system without overheating. A cable's current carrying capacity is determined by its cross-sectional area, and larger cables can handle higher currents. When selecting a cable, ensure its capacity is greater than the maximum current expected in the system.

iii. **Length of the cable run:** The distance between components in the solar system, such as solar panels, charge controllers, batteries, and inverters, influences the cable size selection. Longer cable runs increase the resistance and result in higher voltage drops.

B. Conductor material

Conductor materials are the metallic wires used to conduct electrical energy in cables. The most common conductor materials used in off-grid solar systems are copper and aluminum each with its unique properties and applications.

i. **Copper Cables:** Copper is the most commonly used conductor material in off-grid solar systems due to its excellent electrical conductivity, flexibility, and durability. Copper cables have a lower resistance, which results in lower power losses and higher system efficiency. Additionally, copper cables are more resistant to corrosion, making them suitable for various environments, including humid and coastal areas. However, copper is a more expensive material than aluminum.

ii. **Aluminum Cables:** Aluminum cables are a more cost-effective alternative to copper cables. They are lighter in weight and have a larger diameter for the same current-carrying capacity as copper cables, making them suitable for long cable runs. However, aluminum cables have lower electrical conductivity compared to copper, which can result in higher voltage drops and energy losses. They are also more susceptible to corrosion and are not as flexible.

C. Cable Insulation

Cable insulation is a crucial component of electrical cables, providing a protective barrier between the conducting wire and its surroundings. Insulation prevents electrical shocks, short circuits, and other hazards that can result from exposed conductors. It also helps maintain the integrity of the electrical signal by reducing interference and voltage loss. Several insulation materials are used in electrical cables, each with unique properties and applications, such as Polyvinyl Chloride (PVC), Cross-Linked Polyethylene (XLPE), Ethylene Propylene Rubber (EPR), Polyethylene (PE), and Polytetrafluoroethylene (PTFE).

When selecting cable insulation, it is essential to consider factors like temperature range, UV resistance, moisture resistance, and mechanical durability. The choice of insulation material depends on the specific application, environmental conditions, and system requirements.

12.3 Understanding Wire Gauge Systems

Wire gauge refers to a system used for measuring the diameter of electrical wire. It's a standardized system that assigns a numerical value to the thickness of the wire, with lower numbers representing thicker wires. There are several wire gauge systems used around the world, with the most common ones being the American Wire Gauge (AWG), Standard Wire Gauge (SWG), and International Electrotechnical Commission (IEC) system.

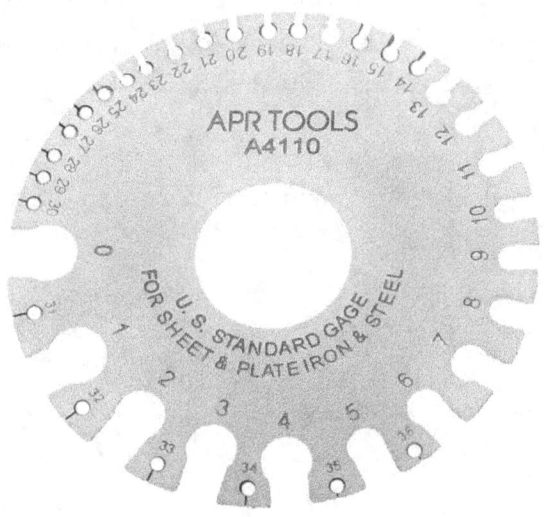

Fig 12.2

Image Credit: APR Tools

i. **American Wire Gauge (AWG)**

    The AWG system is predominantly used in the United States and Canada. In the AWG system, the gauge number is inversely proportional to the wire's diameter, meaning that as the gauge number increases, the wire diameter decreases, and vice versa.

ii. **Standard Wire Gauge (SWG)**

    The SWG system, also known as the British Standard Wire Gauge or Imperial Wire Gauge, is primarily used in the United Kingdom and other countries that were part of the British Empire. The numbering and size increments in the SWG system are different from those in the AWG system.

iii. **International Electrotechnical Commission (IEC) System**

Unlike the AWG and SWG systems, the IEC system measures wire sizes in square millimeters (mm²) of the conductor's cross-sectional area. The IEC system is more straightforward, as the wire size directly corresponds to its cross-sectional area, eliminating the need for gauge numbers.

12.4 Wire Gauge Table

A wire gauge table is an essential reference tool for selecting the appropriate cable size for various electrical applications. It lists wire sizes according to a specific gauge system, typically providing information on wire diameter, cross-sectional area, and resistance per unit length. By consulting a wire gauge table, you can choose the most suitable wire size based on factors such as current-carrying capacity, voltage drop, and power transmission efficiency.

The derated rating is calculated by taking a 25% margin. Derated Ampacity = 1.25 x Max Amperage.

Here's a simplified wire gauge table that includes both copper and aluminum conductors, showing AWG sizes, cross-sectional areas, approximate resistances per unit length, and current capacity:

AWG	Cross-sectional Area (mm²)	Resistance (ohm/km)	Maximum Amperage	Derated Ampacity (A)
18	0.823	39.7	7.5	6.0
16	1.31	25.0	10	8.0
14	2.08	15.8	15	12.0
12	3.31	10.0	20	16.0
10	5.26	6.3	30	24.0
8	8.37	4.0	40	32.0
6	13.3	2.5	55	44.0
4	21.2	1.6	70	56.0
2	33.6	1.0	95	76.0
1	42.4	0.794	110	88.0
1/0	53.5	0.628	125	100.0
2/0	67.4	0.498	145	116.0
3/0	85.0	0.395	165	132.0
4/0	107	0.313	195	156.0
0000	135	0.249	230	184.0

Fig 12.3

Wire Gauge Table

Please note that this table provides approximate values, and actual values may vary depending on the specific type and manufacturer of the cable. The table does not include factors such as temperature and installation conditions, which can also influence cable performance.

How to Use a Wire Gauge Table:

1. Find a wire size in the AWG table that matches your system's needs, considering factors like current carrying capacity and voltage drop. The table will show wire sizes, diameters, cross-sectional areas, and resistances per unit length (ohms per 1000 feet or ohms per kilometre).

2. Compare wire sizes: If choosing between two wire sizes, think about the differences in cost, energy efficiency, and installation ease. Bigger wire sizes usually have less voltage drop and better efficiency but might be more costly and harder to install.

3. Review manufacturer's recommendations: Check the cable manufacturer's guidelines to make sure the selected wire size is suitable for your specific project, as actual values can vary based on the cable type and manufacturing process.

12.5 Calculating Appropriate Cable Size

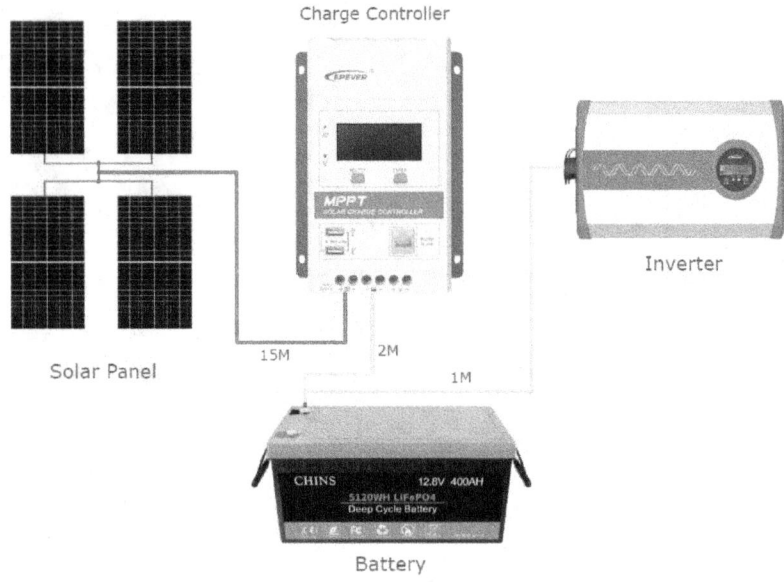

Fig 12.4

Solar DC Cable

You can find out the correct size of cable required for your application either by using an Online Calculator or using the following manual method.

Let's go through an example calculation for an off-grid solar PV system. We will size the cables connecting the solar panels to the charge controller, charge controller to the battery bank, and battery bank to the inverter.

Assumptions:

- 4 solar panels, each with 540W power output, Imp = 13.96A, Vmp = 41.7V, Isc = 13.64A, Voc = 49.5V

- Panels are connected in 2 strings of 2 panels each (series-parallel configuration) 48V battery bank with a capacity of 400Ah
- MPPT charge controller with a maximum input current of 40A
- 48V inverter with a maximum input current of 100A
- Cable lengths: 15m (solar panels to charge controller), 2m (charge controller to battery bank), 1m (battery bank to inverter)
- Allowable voltage drop: 3%

Step 1: Determine the total current

Total power of the solar array (two strings of two panels each):

4 panels * 540W = 2160W

Voltage of one string (two panels in series): Vmp = 41.7V x 2 = 83.4V

Current of one string (two panels in parallel): Imp = 13.96A x 2 = 25.92A

Step 2: Calculate the wire resistance

Wire resistance can be calculated by using Ohm's Law (R = V/I)

Resistance per kilometer (R/km) = R / Cable length in km

- **Solar panel to charge controller (15m):**
  Voltage drop allowed (3%) = 0.03 x 83.4V = 2.502V
  R = 2.502V / 25.92A = 0.0965 ohms
  Resistance per kilometer = 0.0965 ohms x 1000 /0. 015km = 6.43 ohms/km
- **Charge Controller to Battery Bank (2m):**
  Battery bank voltage: 48V and Maximum charge current: 40A (charge controller)
  Voltage drop = (3% of 48V) = 0.03 x 48V = 1.44V
  R = 1.44V / 40A = 0.036 ohms
  Resistance per kilometer = 0.036 ohms / 0.002 km = 18 ohms/km
- **Battery Bank to Inverter (1m):**
  Inverter input voltage: 48V and Maximum input current: 100A
  Voltage drop = (3% of 48V) = 0.02 x 48V = 1.44V
  R = 1.44V / 100A = 0.0144 ohms
  Resistance per kilometer = 0.0144 ohms / 0.001 km = 14.4 ohms/km

Step 3: Determine the wire gauge using the AWG table

Find the AWG value with a resistance closest to or lower than the calculated resistance per kilometer for each segment.

- **Solar panel to charge controller (6.43 ohms/km):**
  From the AWG table, select a copper cable with resistance <= 6.43 ohms/km and derated amperage >= 25.92A. A suitable choice would be AWG 8, with a resistance of 4 ohms/km and adjusted amperage of 32A.

- **Charge controller to battery bank (18 ohms/km):**
  From the AWG table, select a copper cable with resistance <= 18 ohms/km and adjusted amperage >= 40A. A suitable choice would be AWG 6, with a resistance of 2.5 ohms/km and derated amperage of 44A.

- **Battery bank to inverter (14.4 ohms/km):**
  From the AWG table, select a copper cable with resistance <= 14.4 ohms/km and an adjusted ampacity >= 100A. A suitable choice would be 1/0 AWG, with a resistance of 0.628 ohms/km and a derated ampacity of 100A.

12.6 DC Cable Size Chart

You may also refer this Blue Sea Cable Sizing chart for quickly determine the DC cable size.

For more details on how to use this chart, you can refer Blue Sea official website

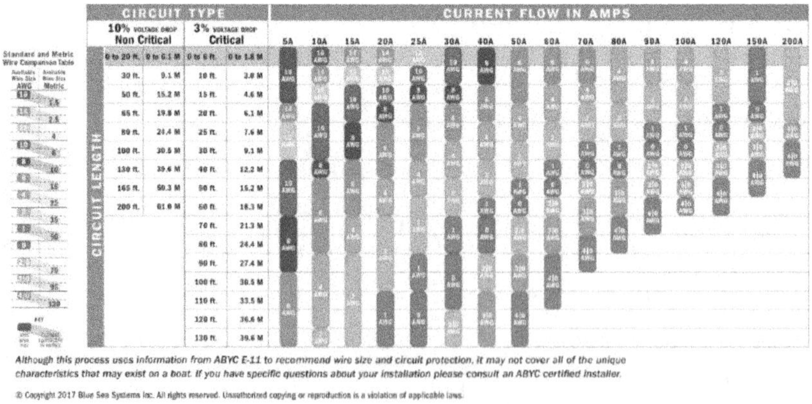

Fig 12.5

Credit: Blue Sea Systems Inc

Chapter 13
DIY Off Grid Solar Power Installation

The prices of solar panels have been falling gradually but the cost of an off-grid solar system setup is rising steadily. However, anyone with basic knowledge of Electricity and a toolbox can install it on their own. This will reduce the overall system cost substantially and you will learn a lot.

Embarking on a DIY installation can be a rewarding project, but it requires careful planning, and adherence to safety practices. Here is a systematic guide to help you through this process:

13.1 Site Assessment

It involves evaluating the location to ensure it is suitable for solar installation and to optimize the system's performance. The detailed approach to conducting a site assessment:

Sunlight Assessment:

Conducting a sunlight assessment is a crucial step in the planning and installation of a solar power system, as it evaluates the potential solar energy generation of a specific location. You have to determine the average solar insolation (sunlight intensity) for your location. This information, usually measured in kWh/m²/day, is available from solar radiation databases or local meteorological data.

The next important task is to performing a thorough shading analysis. This involves identifying potential obstacles such as trees, buildings, and other structures that may cast shadows over the installation site at various times throughout the day and year. Tools like solar pathfinders or digital shading analysis software are instrumental in this process, offering detailed insights into how these obstacles affect sunlight exposure.

Space Availability:

When designing off-grid solar systems for homes, RVs, and boats, space availability is a critical factor that significantly influences the system's design and capacity.

- **For Off-Grid Homes:**

In homes, rooftop space is often the primary location for solar panel installation. The size and layout of the roof determine how many panels can be installed, with considerations for obstructions like chimneys or vents. If rooftop space is inadequate or unsuitable due to shading, ground-mounted systems offer an alternative, though they require clear, unshaded land. The balance between the available space and the home's energy needs is essential to ensure sufficient power generation.

- **For RVs:**

For RVs, space constraints are more pronounced. The limited roof area, often shared with other equipment, restricts the number of panels that can be installed. To augment this, portable solar panels are a practical solution, providing additional capacity when the RV is stationary. Given these space limitations, opting for high-efficiency solar panels is advantageous as they deliver more power per square foot.

- **For Boats:**

Boats present unique challenges, with solar panels typically installed on decks, bimini tops, or railings. The harsh marine environment necessitates robust, marine-grade panels that are resistant to saltwater and waterproof. Flexible solar panels are particularly useful in maximizing limited space on boats, as they can conform to various surfaces and shapes

13.2 Planning and Design

Assess Energy Needs: Calculate the total daily energy consumption of your household. List all electrical appliances and their power consumption in watts, considering how many hours a day each is used.

System Sizing: Based on your energy needs, determine the size of the solar panel array, the capacity of the battery bank, and the specifications for the charge controller and inverter.

Component Selection: Choose solar panels, batteries (such as lead-acid or lithium-ion), a charge controller (MPPT or PWM), and an inverter (pure sine wave or modified sine wave) that fit your system's requirements.

13.3 Gathering Materials

Purchase Components: Acquire all necessary components, including solar panels, batteries, a charge controller, an inverter, cables, connectors, mounting brackets, and safety gear.

Tools: Ensure you have the necessary tools for installation, such as a drill, screwdrivers, wrenches, wire strippers, and a multimeter.

13.4 Mounting Solar Panels

Before beginning the actual installation, the site itself must be adequately prepared. This involves ensuring the roof's structural integrity for rooftop installations or clearing and levelling land for ground-mounted systems. In the case of roof installations, a thorough inspection of the roof is necessary to identify the best attachment points and to ensure that the roof can bear the weight of the solar panels and mounting equipment. For ground mounts, choosing a location with optimal sun exposure and minimal shading throughout the day is crucial. The area should be cleared of debris, vegetation, or other obstructions, and the ground levelled to provide a stable base for the mounting structure.

Once the site is prepared, the next step is to plan the layout of the solar panels and the routing of electrical wiring. This includes determining the best arrangement of panels to optimize sun exposure and space usage, as well as planning the path of wires from the panels to the inverter and batteries connection. Careful planning of the wiring route is essential to minimize cable length and potential energy loss, as well as to ensure safety and aesthetic appeal. Ensure that wiring is protected from environmental elements and secured against physical damage.

Once mounted, inspect the installation to ensure everything is secure. Test the panels to ensure they are functioning correctly and providing the expected output.

The next task is to adjust the panels orientation and tilt angle to maximize the energy production.

In the Northern Hemisphere, the ideal orientation is true south, while in the Southern Hemisphere, panels should face true north. This maximizes exposure to sunlight throughout the day. However, practical constraints, such as the layout of a roof on which panels are mounted, might necessitate deviation from this ideal. Even so, east- or west-facing panels can still be effective, particularly if energy demand aligns with sunrise or sunset times.

The tilt angle is typically set to *approximate the geographical latitude* of the installation site. This positioning is generally considered optimal for year-round energy production. However, for enhanced efficiency, some systems employ adjustable mounts that allow the tilt angle to be altered seasonally – decreased during summer and increased in winter to align with the sun's lower trajectory.

13.5 Battery Bank Setup

Setting up a battery bank is a critical process that involves selecting the right type and number of batteries and configuring them correctly to store the energy generated by your solar panels.

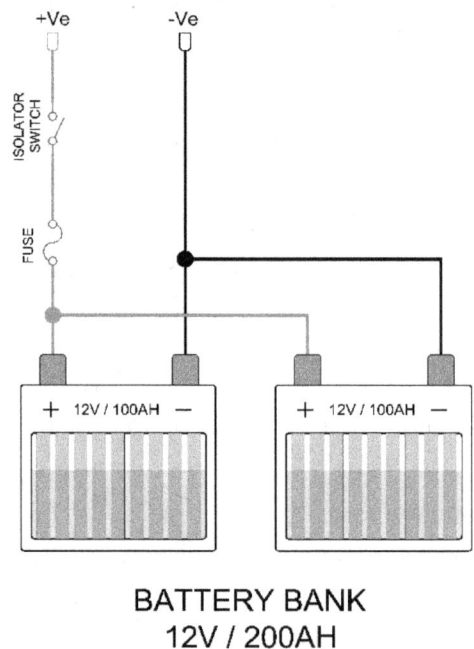

BATTERY BANK
12V / 200AH

Fig 13.1

For battery configuration, you may review the section related to series and parallel connections in the battery section of this book before wiring your batteries.

When making connections in the battery bank, it is essential to use appropriate cables that are capable of handling the system's current and voltage. The cables should be of a suitable gauge to prevent significant voltage drops and ensure efficient power transmission. It is advisable to check the manual of the charge controller for the recommended cable size.

Secure connections are crucial; all terminals must be tightened properly to avoid any loose connections, which can lead to arcing or overheating, posing safety risks. Additionally, always double-check the polarity when making connections. Incorrect polarity can cause short circuits, potentially damaging the batteries and other system components.

Install an appropriate fuse or circuit breaker on the positive cable between the battery bank and the inverter/charge controller for overcurrent protection.

13.6 Installing Charge Controller

The first step in the installation process is to select an appropriate location for the charge controller. It should be easily accessible for monitoring and maintenance, well ventilated to dissipate heat effectively, and close to the battery bank to minimize voltage drop. Additionally, the chosen location should be dry, shielded from direct sunlight, and protected from extreme temperatures and water.

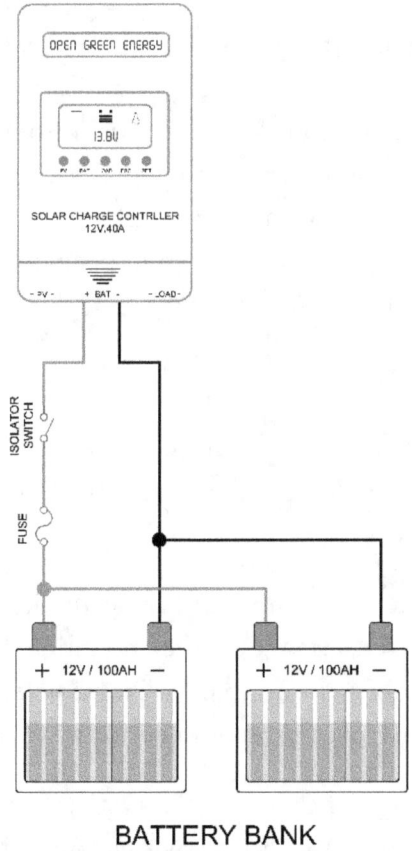

BATTERY BANK
12V / 200AH

Fig 13.2

Once a suitable location is identified, securely mount the charge controller using screws or mounting brackets, ensuring it is stable and referencing the manufacturer's installation manual for specific guidelines.

Wiring Connections:

1. Connect the charge controller to the battery bank first. Connect the positive and negative wires from the controller to the corresponding terminals of the battery bank.
2. Install a fuse or circuit breaker between the battery bank and the charge controller for safety.
3. Next, connect the solar panel(s) to the charge controller.
4. If the charge controller has a load terminal, connect your DC loads to these terminals.

Note: Always double-check the polarity of the connections. Reversing the polarity can cause serious damage to the charge controller and the solar system.

After completing the wiring, you will notice that the display on the charge controller will become illuminated. At this stage, it is important to specify the type of battery your system is using. To correctly configure this, refer to the manufacturer's manual provided with your charge controller. The manual will have detailed instructions on how to select the battery type, ensuring that the charge controller select its appropriate charging algorithm and parameters to match the specific requirements of your battery, whether it's lead-acid, AGM, lithium-ion type.

13.7 Inverter Installation

Similar to charge controller, first you have to select an appropriate location for installation. Place the inverter close to the battery bank to minimize voltage drop and energy loss.

When mounting the inverter, it is important to secure it on a stable, vertical surface, such as a wall or a shelf, ensuring that the mounting arrangement can withstand vibrations, especially in mobile environments like vehicles or boats. Always refer the manufacturer instruction manual for recommended installation procedure.

When wiring an inverter for your off-grid solar system, you have two main options for connecting it to your battery bank: directly wiring to the battery terminals or using a busbar.

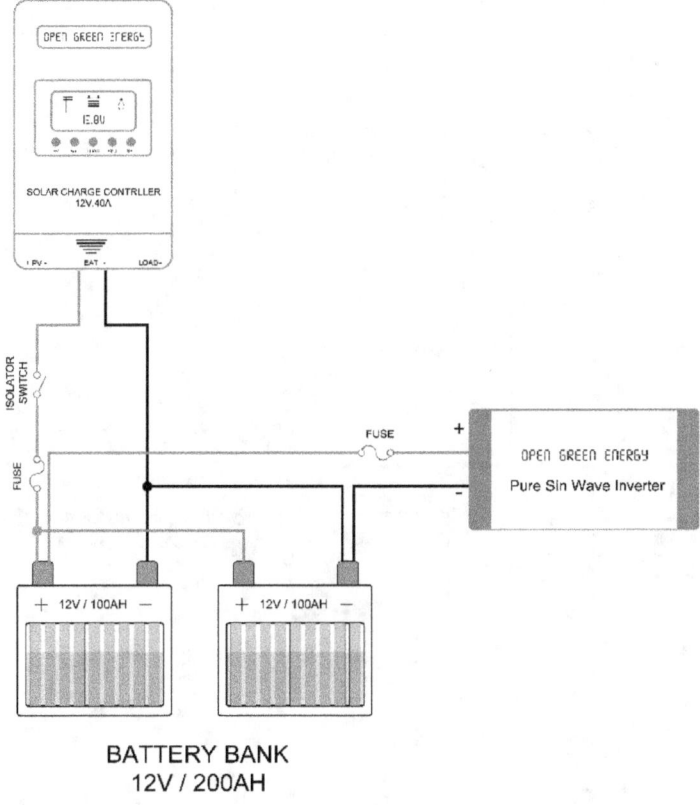

Fig 13.3: Direct Wiring

Wiring directly to the battery terminals is a straightforward approach where you connect the inverter's positive and negative cables straight to the respective terminals on the battery bank. This method is simple and often used in smaller systems.

Alternatively, you can opt to wire the inverter via a busbar. A busbar serves as a centralized point for connecting multiple components, not just the inverter but also the charge controller and any other systems linked to the battery bank. This setup is neat and organized, reducing wiring clutter, and is particularly advantageous in larger systems. It makes future expansions or additions easier and ensures a more even distribution of power and charging across the battery bank.

You may refer the cable and fuse sizing section of this book for selecting the correct cable and fuse/breaker for battery bank to inverter.

Attention must be paid to connect the positive and negative leads correctly to their respective terminals on the battery to prevent damage.

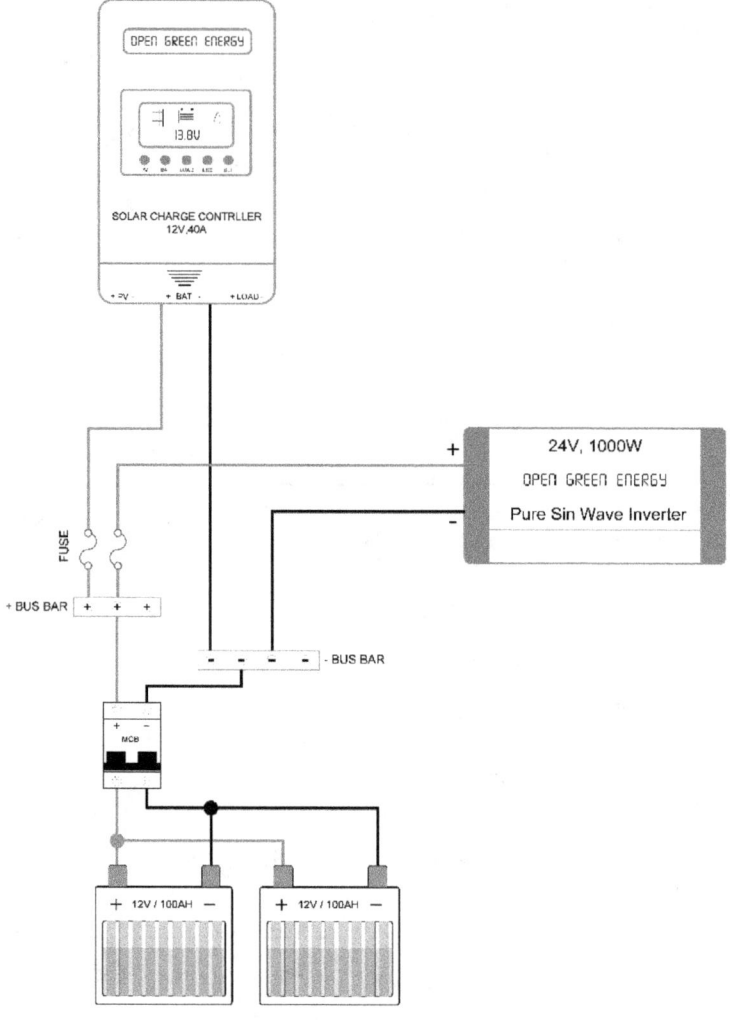

Fig 13.4: Wiring using a Busbar

Connect the inverter AC output terminal to the AC distribution panel or directly to AC appliances. All AC connections should comply with electrical standards and safety regulations.

Properly ground the inverter following local electrical codes and the manufacturer's recommendations.

13.8 DC Fuse Box Installation

Securely mount the fuse box to a wall or a stable surface, ensuring it is well-organized and clearly labelled.

You can connect the DC fuse box directly to the load terminals of the charge controller. The positive and negative cable from the charge controller's load terminal should run to the fuse box positive and negative input terminals respectively.

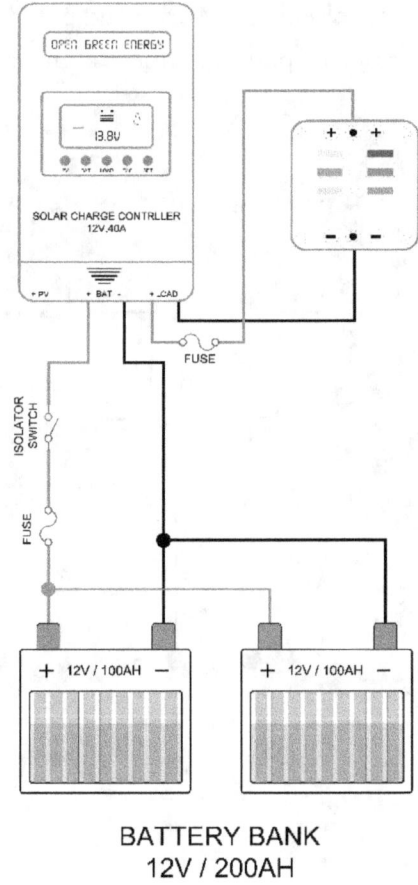

BATTERY BANK
12V / 200AH

Fig 13.5

However, in most of the charge controller's load terminal are limited to 20A. That means you can connect the maximum DC loads up to 240W (12V x 20A =240W).

If you have to connect many DC loads, it is recommended to connect the DC fuse box to the battery terminals through a busbar.

13.9 Wiring the Solar Panels

The wiring configuration largely depends on the voltage requirements of your battery bank and the input specifications of your charge controller.

First, determine whether a series or parallel connection is more appropriate. A series connection is suitable for systems with higher voltage battery banks or charge controllers. On the other hand, a parallel connection suitable for systems to increase the current while maintaining the panel voltage. In cases where high voltage and high current are both required, a combination of series and parallel wiring is employed. This involves creating multiple strings of panels connected in series, then connecting these strings in parallel.

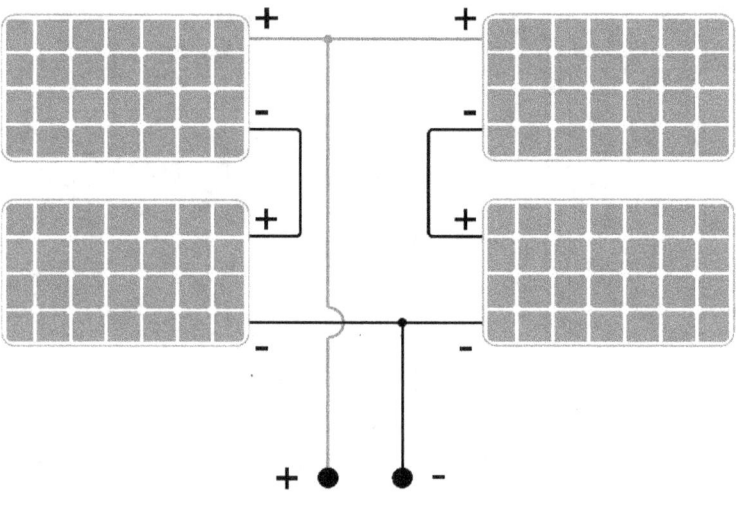

Fig 13.6

Run the positive and negative leads from the solar array to the charge controller, ensuring to include an in-line fuse or breaker for safety. It is vital to maintain correct polarity when connecting these leads to the charge controller to prevent damage. Proper grounding of the solar array and the charge controller is also crucial for safety, adhering to local electrical codes.

13.10 Final Testing

After installing your off-grid solar system, it is important to test it to make sure everything is working properly. A simplified guide on how to do that is given below:

- **Look over Everything:** First, take a look at all the connections. Make sure all wires are securely connected and there is no damage to any parts.

- **Check the Voltages:** Use a multimeter to check the voltage from your solar panels, battery bank, and the output from your charge controller. You want to make sure these numbers match what you expect.

- **Test the Solar Panels:** Measure how much current and power your solar panels are producing on a sunny day. This tells you if they are working, as they should.

- **Use Some Appliances:** Turn on some lights or appliances connected to the system. Watch to see if the inverter is handling the load and the voltage remains stable.

- **Charge Controller Check:** Look at the charge controller's display to make sure it is going through the correct charging stages – like bulk, absorption, and float.

- **Inverter Output:** Check the AC output from your inverter with a multimeter. It should be giving a stable and correct voltage for your appliances.

- **Safety Tests:** Make sure any fuses or circuit breakers in your system trip off correctly if there is an overload. If your system has ground fault protection, test this too.

- **Monitor the System:** If your system has a way to monitor performance, like a remote monitor, check that it is showing the right information about voltage, current, and battery charge.

13.11 Example Wiring Diagrams

1.   500W/12V Off-Grid Solar System

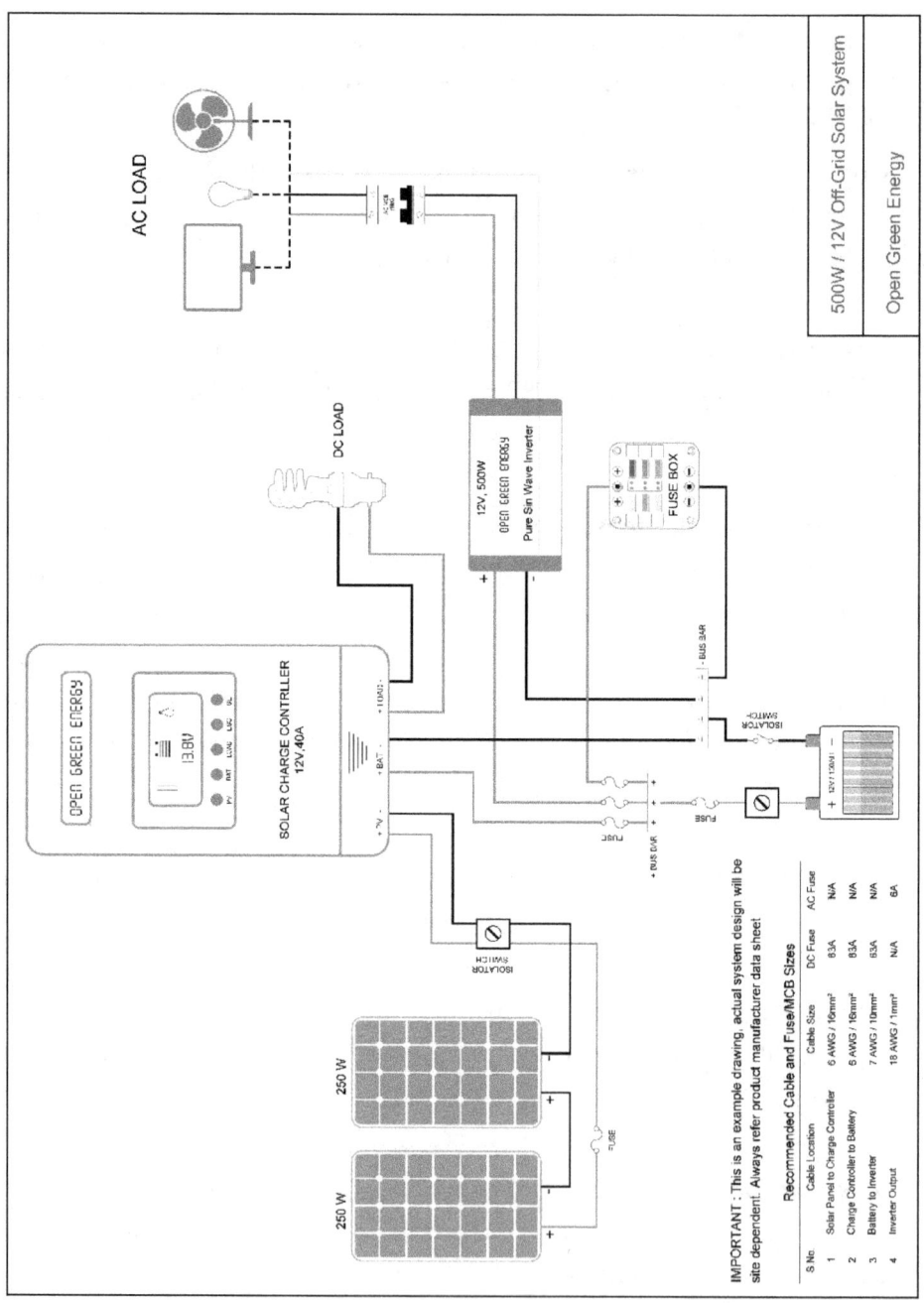

Fig 13.7

2. 1000W/24V Off-Grid Solar System

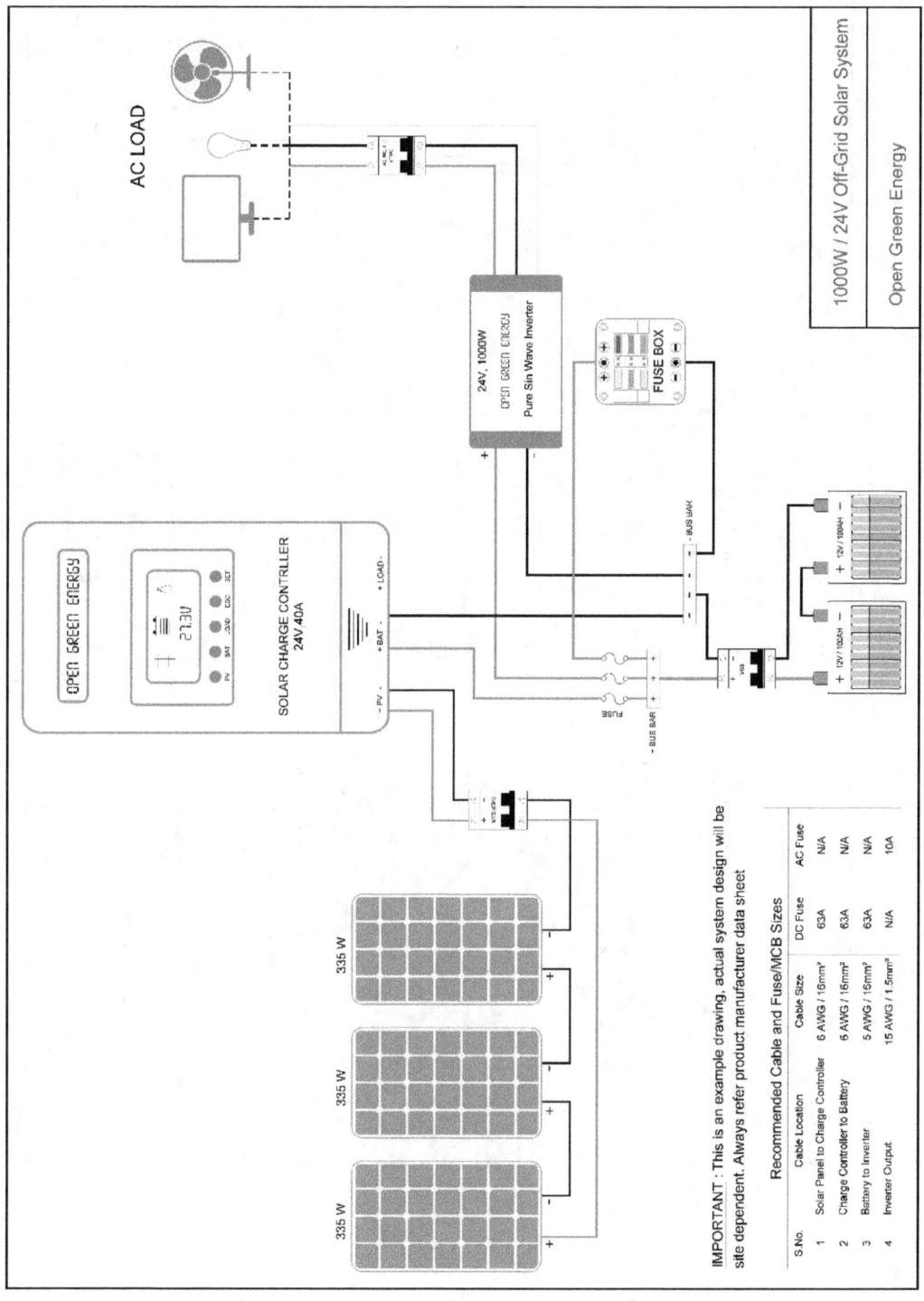

Fig 13.8

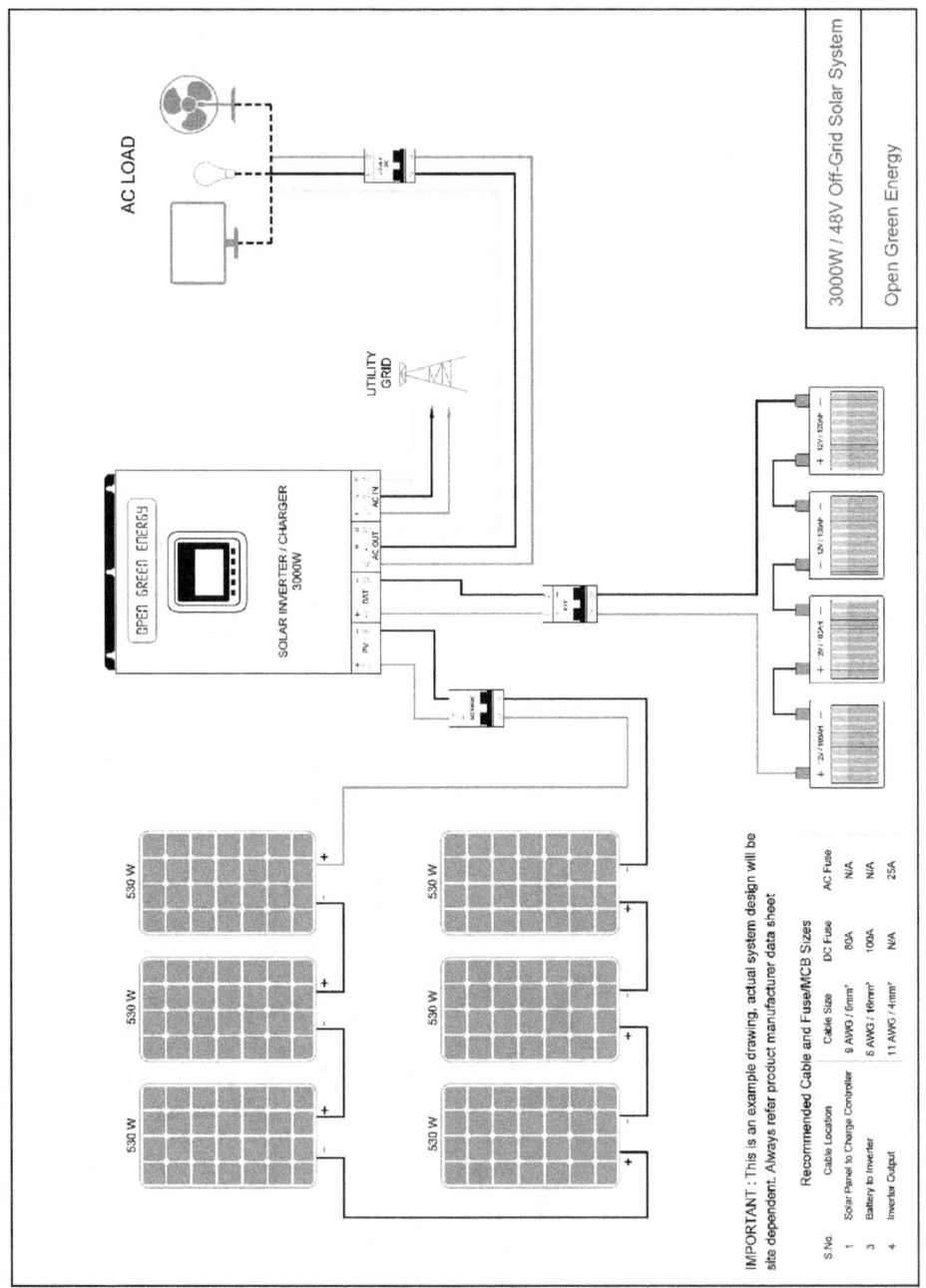

Fig 13.9

You can view these drawings in high resolution and color on the website
https://www.opengreenenergy.com/ebook01 (Password: **OGE_SOL@23**)

Chapter 14
Maintenance and Care

14.1 Solar Panels

Proper maintenance and care are essential to keep your solar panels operating efficiently and extend their lifespan. Regular cleaning and inspection of your solar panels can help you get the most out of your off-grid solar system. In this section, we will discuss essential maintenance tips and best practices for solar panel care.

14.1

Source: Shoalhaven Solar

1. Cleaning Solar Panels:

- Keep the solar panels free of dust, dirt, and debris, which can accumulate and reduce the panels' efficiency. Clean the panels regularly, depending on your location and environmental factors (such as dust or pollen levels). In general, cleaning the panels every 3-6 months is recommended. However, you can get maximum gain if the cleaning is done in a month.

- Use a soft brush or sponge and mild, soapy water to clean the panels gently. Avoid using abrasive materials or harsh chemicals, as they may damage the glass surface.

- Clean the panels during the early morning or late afternoon, when the panels are cooler, to prevent thermal shock and potential damage.

Fig 14.2

2. Inspecting Solar Panels:

- Regularly inspect the solar panels for any signs of damage, such as cracks, discoloration, or delamination. These issues may affect the panels' performance and may require repair or replacement.

- Check the mounting hardware and racking system for signs of corrosion, loose bolts, or other issues that may compromise the structural integrity of the solar array.

- Ensure that the wiring and connections are secure and free of corrosion or damage. If you notice any issues, consult a professional to address them.

Fig 14.3

Source: Canstar Blue

196

3. Monitoring System Performance:

- Keep track of your solar panels' output and overall system performance to identify any potential issues or decreases in efficiency. Regularly monitoring your system can help you detect problems early and address them before they become more severe.

- Use a solar monitoring system or consult your inverter's display to access real-time data about your system's performance.

4. Preventative Maintenance:

- Trim any overgrown trees or vegetation around your solar panels to prevent shading, which can significantly reduce the panels' efficiency.

- Ensure proper ventilation and airflow around the solar panels to prevent overheating and maintain optimal performance.

- Adhere to the manufacturer's recommended maintenance schedule and guidelines for your specific solar panel type.

14.2 Battery Bank

Proper maintenance and care are crucial for extending the life of your battery storage system, ensuring optimal performance and reducing long-term costs. In this section, we will discuss essential maintenance tips and best practices for different battery types to help you get the most out of your off-grid solar system.

1. Flooded Lead-Acid (FLA) Batteries:

- Regularly check electrolyte levels and add distilled water as needed to keep the plates submerged.

- Clean the battery terminals to prevent corrosion and ensure good electrical connections.

- Equalize the battery once a month by charging it at a higher voltage for a short period to balance the cells and prevent sulfation.

2. Sealed Lead-Acid (SLA) Batteries:

- Keep the batteries clean and free of dust and debris.

- Monitor the battery voltage and temperature to ensure they stay within the recommended ranges.

- Avoid overcharging and over-discharging the batteries, which can lead to a reduced lifespan.

3. Lithium-Ion and LiFePO4 Batteries:

- Ensure the battery management system (BMS) is functioning correctly to protect the battery from overcharging, over-discharging, and high temperatures.

- Keep the batteries in a temperature-controlled environment, as extreme temperatures can negatively impact their performance and lifespan.

- Store the batteries at a partial state of charge (typically 50-60%) if not in use for an extended period.

4. General Battery Maintenance Tips:

- Regularly inspect the battery enclosure for any signs of damage or wear, ensuring proper ventilation and airflow.

- Keep a log of battery performance, including voltage, temperature, and capacity, to track the health of your battery system and identify potential issues.

- Adhere to the manufacturer's recommended maintenance schedule and guidelines for your specific battery type.

By following these maintenance tips and best practices, you can extend the life of your battery storage system, improve its performance, and reduce long-term costs associated with battery replacement and system downtime. Regular maintenance is essential for all battery types, but it is particularly crucial for lead-acid batteries, which have higher maintenance requirements than lithium-ion and LiFePO4 batteries

14.3 Inverter and Charge Controller

Proper maintenance and care are crucial for ensuring the efficient operation and longevity of your off-grid solar system's components, including inverters and charge controllers. In this section, we will discuss essential maintenance tips and best practices for component care.

1. Keep the Components Clean and Dust-Free:

- Regularly clean the exterior and air vents of system components with a soft, dry cloth or brush to remove dust and debris. Dust accumulation can lead to overheating and reduced efficiency.

- Avoid using water, solvents, or other liquids to clean the components, as they may cause damage to the electrical parts.

2. Inspecting the Components:

- Regularly inspect the system components for any signs of damage or wear, such as loose connections, frayed wires, or corrosion. If you notice any issues, consult a professional for repair or replacement.

- Ensure that the components are securely mounted and that the mounting hardware is in good condition. Check for any signs of corrosion, loose bolts, or other issues that may compromise the components' stability.

3. Monitoring Component Performance:

- Keep track of the performance of system components, including output power, efficiency, and error codes. Regular monitoring can help you detect issues early and address them before they become more severe.

- Consult the user manuals for the components for information on how to access performance data, either through the component's display or a remote monitoring system.

4. Ensure Proper Ventilation and Cooling:

- Make sure the components are installed in well-ventilated areas with sufficient clearance on all sides to allow for proper airflow and cooling. Overheating can reduce their efficiency and lifespan.

- Avoid installing the components in direct sunlight or near heat-generating equipment, as this can contribute to overheating.

5. Preventative Maintenance:

- Regularly check for firmware updates, if applicable, and follow the manufacturer's recommendations for upgrades. Updated firmware can help improve performance and address potential issues.

- Adhere to the manufacturer's recommended maintenance schedule and guidelines for the specific components in your system.

- If you experience recurring issues or notice a decline in performance, consult a professional for an in-depth inspection, maintenance, or potential replacement.

14.4 Balance of System (BOS) Components

The Balance of System (BOS) components play a vital role in the efficient operation and longevity of your off-grid solar system. BOS components include mounting structures, wiring, and other electrical components necessary for the solar system to function. In this section, we will discuss essential maintenance tips and best practices for BOS component care.

1. Inspect and Maintain Mounting Structures:

- Regularly check mounting structures for signs of wear, corrosion, or loose bolts. Ensure that the mounting hardware is in good condition and securely fastened.

- Keep the mounting structures clean and free from dirt or debris, as accumulated dirt can lead to premature wear and corrosion.

2. Check and Maintain Wiring and Electrical Connections:

- Regularly inspect the wiring and electrical connections for signs of wear, corrosion, or damage. Loose or damaged connections can lead to inefficiencies, energy loss, or even safety hazards.

- Tighten any loose connections and replace damaged wires or connectors as needed.

- Ensure that all wiring is adequately secured and protected from harsh weather conditions, such as direct sunlight, rain, or snow.

3. Maintain and Protect Combiner Boxes and Disconnects:

- Regularly check the combiner boxes and disconnect switches for signs of wear, corrosion, or damage. Ensure that the enclosures are in good condition and securely fastened.

- Keep these components clean and free from dirt or debris to prevent overheating and ensure proper functioning.

4. Inspect and Maintain Grounding System:

- Regularly check the grounding system for proper connections and signs of wear, corrosion, or damage. A well-maintained grounding system is essential for the safety and stability of your off-grid solar system.

- Tighten any loose connections and replace damaged components as needed.

5. Monitor System Performance:

- Regularly monitor the performance of your off-grid solar system, including energy production, efficiency, and potential issues related to BOS components. Early detection of issues can help address them before they escalate.

- Consult the user manuals for your specific components for information on performance monitoring and recommended maintenance schedules.

Proper maintenance and care for all components of your off-grid solar system are essential to ensure efficient operation, long-term performance, and reliability. Regularly inspect and clean solar panels, inverters, charge controllers, batteries, and balance of system components to prevent issues related to wear, corrosion, or damage. Monitoring system performance and adhering to the manufacturer's recommended maintenance schedules will help detect potential problems early and address them before they escalate. By taking a proactive approach to maintaining your solar energy system, you can optimize its performance, extend the lifespan of its components, and maximize your return on investment.

Chapter 15
Cost and Financing Options

Embarking on the journey towards adopting off-grid solar energy systems involves several considerations, and understanding the costs and financing options is crucial. The affordability of such systems plays a significant role in the decision-making process, as it directly impacts your return on investment. Factors such as component costs, installation, maintenance, and replacement expenses, as well as system size and location, all influence the overall cost of an off-grid solar system. This chapter aims to provide a comprehensive overview of these factors and shed light on the various financing options available. We will also discuss the pros and cons of DIY versus professional installation and explore tips for reducing costs. By the end of this chapter, you will have the necessary knowledge to make informed decisions that maximize the benefits of solar energy for your unique situation.

15.1 Cost Factors for Off-Grid Solar System Installation

When considering an off-grid solar system, it is essential to estimate the various costs involved. This will help you make informed decisions about the size and components of your system, as well as the best financing options.

1. **System Size**

- **Energy Needs:** The total cost is heavily influenced by how much electricity you need to generate. This depends on your daily energy usage, which is calculated in kilowatt-hours (kWh). More energy usage means more solar panels and a larger battery bank, leading to higher costs.

- **Peak Sunlight Hours:** Your geographical location affects how much sunlight your solar panels can harness. Areas with fewer peak sunlight hours may require more or larger panels to generate the same amount of electricity, impacting costs.

- **Redundancy and Over-sizing:** To ensure reliability, especially during periods of low sunlight or higher than usual consumption, systems are often oversized, which increases costs.

2. Component Quality

- **Solar Panels:** Higher efficiency panels cost more but generate more electricity per square foot. They're beneficial if space is a constraint or if you want to maximize energy production.
- Batteries: The type (lead-acid vs. lithium-ion) and capacity of batteries greatly affect cost. Lithium-ion batteries are more expensive but have a longer lifespan and better depth of discharge (DoD).
- **Inverters and Charge Controllers:** Higher quality inverters and charge controllers that offer more efficiency and features (like MPPT charge controllers) come at a higher price.

3. Installation Complexity

- **Site Accessibility:** Difficult-to-access sites can increase installation costs due to transportation challenges and the extra time and labor required.
- **Mounting System:** Standard rooftop mounts are generally less expensive than ground mounts or pole mounts. Tracking systems, which follow the sun, are more efficient but much more costly.
- **System Integration:** Integrating the solar system with existing structures or systems can add to the complexity and cost, especially if custom solutions are required.

4. Additional Features

- **Backup Generator:** Including a diesel or gas generator for additional backup during prolonged low sunlight periods adds to the cost.
- **Energy Management System:** Advanced systems for monitoring and managing energy use can improve efficiency but also add to the initial investment.
- **Smart Connectivity:** Features like remote monitoring, smart home integration, or Internet of Things (IoT) compatibility increase costs.

5. Labor Costs

- **Professional Installation:** The cost of labor varies significantly based on location and the expertise of the installer. Professional installation ensures safety and system efficiency but adds to the overall cost.

- **Permits and Inspections**: The cost and necessity of permits and inspections vary by location but are often required and can add to the total cost.

6. **Other Factors**

- **Shipping Costs:** For remote locations, the cost of shipping heavy and bulky components like solar panels and batteries can be significant.
- **Maintenance and Replacement Costs:** Long-term costs include maintenance, repairs, and eventual replacement of components, particularly batteries.
- **Insurance and Warranty Extensions:** Protecting your investment with additional insurance or purchasing extended warranties can also add to the upfront cost.

15.2 Financing Options

When it comes to financing an off-grid solar system, there are several options available to help you manage the initial costs. Here are some common financing methods you might consider:

1. Personal Savings:

Using your personal savings can be the most straightforward way to finance your off-grid solar system. This option allows you to avoid any interest or loan fees and provides you with full ownership of the system from the beginning.

2. Loans:

You may consider taking out a loan from a bank or other financial institution to finance your solar system. This could be a personal loan, home equity loan, or even a specific solar loan. Interest rates and terms will vary depending on the institution and your credit history. Keep in mind that loans will add to the overall cost of your system due to interest payments.

3. Government Grants and Incentives:

Some countries or states offer grants, tax credits, or other incentives for installing renewable energy systems. These incentives can help offset the initial costs of your solar system. Research your local area to find out if any programs are available and if you qualify for them.

4. Lease or Power Purchase Agreement (PPA):

Some solar companies offer leasing or power purchase agreements (PPAs) for off-grid systems. With a lease, you pay a fixed monthly fee to the solar company for the use of their equipment. In a PPA, you agree to purchase the energy generated by the solar system at a predetermined rate. Both options can help reduce upfront costs but may result in higher overall costs in the long run.

5. Crowd-funding or Peer-to-Peer Lending:

Some individuals have turned to crowd-funding platforms or peer-to-peer lending to finance their off-grid solar projects. These options can provide access to funds but may come with higher interest rates or fees compared to traditional financing methods.

When exploring financing options, it is essential to weigh the pros and cons of each method and consider the overall impact on the cost of your off-grid solar system. Be sure to research and compare various options to find the best fit for your specific situation and needs.

15.3 Evaluating Return on Investment (ROI)

One of the critical factors to consider when investing in an off-grid solar system is the return on investment (ROI). ROI helps you determine how long it will take for your solar system to pay for itself through the savings on electricity bills and other associated benefits. To evaluate the ROI of your off-grid solar system, you will need to take several factors into account:

1. System Cost: Consider the total cost of your solar system, including equipment, installation, permits, and any financing interest or fees.

2. Energy Savings: Calculate the amount of money you'll save on electricity bills by switching to solar energy. You can estimate this by comparing your current energy consumption and costs with the expected output and performance of your solar system.

3. Maintenance and Operating Costs: Factor in the ongoing costs of maintaining and operating your solar system, such as cleaning, repairs, and battery replacement.

4. Incentives and Rebates: Account for any financial benefits you'll receive from government grants, tax credits, or other incentives.

5. System Lifespan: Consider the expected lifespan of your solar system, typically around 25-30 years for solar panels and 10-15 years for batteries. Keep in mind that some components may need to be replaced during the system's life.

Once you have gathered all the necessary information, you can calculate the ROI by dividing the net benefits (total savings and incentives) by the initial investment (system cost). This will give you a percentage that represents the return on investment. The higher the percentage, the better the ROI.

15.4 Example Calculation

Let us create a practical example by outlining a scenario where a homeowner is looking to install an off-grid solar system. We will explore different costs involved and discuss potential financing options.

Example Calculation for Installing an Off-Grid Solar System for a Small Home

Homeowner's Need: The homeowner wants to power a small, energy-efficient home located in a rural area. The home requires a system capable of generating an average of 5 kWh per day.

Cost Breakdown

I. **Components Cost:**
- Solar Panels: 3 kW system to meet daily energy needs considering average sunlight hours. Estimated cost: $6,000.
- Battery Storage: Lithium-ion battery bank for energy storage, able to store 10 kWh (2 days of autonomy). Estimated cost: $7,000.
- Inverter: 3 kW inverter. Estimated cost: $2,000.
- Charge Controller: MPPT charge controller. Estimated cost: $500.
- Mounting Hardware, Cables, and Miscellaneous: Estimated cost: $1,500.
- Total Components Cost = $7000+$2000+$500+$1500 = $17000

II. **Installation Costs:**
- Labour and Professional Installation: Estimated cost: $3,000.
- Permits and Inspections: Estimated cost: $500.
- Total Installation Cost = $3000+$500 = $3500

Total Estimated System Cost = Components Cost + Installation Costs

$$= \$17000 + \$3500 = \$20,500$$

Financing Options

I. Personal Savings

If the homeowner has enough savings, paying upfront could be the most cost-effective option.

II. Green Energy Loan

- Assume the homeowner opts for a specialized solar loan.
- Loan Amount: $20,500
- Interest Rate: 4.5% per annum
- Loan Term: 10 years
- Monthly Payment: Approximately $212
- Total Interest Paid Over 10 Years: Approximately $5,440

III. Government Grants and Tax Incentives

- Let us say there is a government incentive offering a 30% tax credit.
- Initial System Cost: $20,500
- Tax Credit: $6,150
- Adjusted System Cost After Tax Credit: $14,350
- If combined with a loan, this reduces the loan amount or can be received as a tax rebate.

IV. Home Equity Loan (If Applicable)

If the homeowner has equity in their property, they might opt for a home equity loan with potentially lower interest rates.

V. Long-term Considerations and ROI

- **Electricity Cost Savings:** By generating their own electricity, the homeowner saves on utility bills, which can be significant over the years.
- **Maintenance Costs:** Maintenance and potential replacement costs, especially for the battery system, should be factored in.
- **Property Value Increase:** Solar systems can increase property value, a benefit if the homeowner decides to sell in the future.

Conclusion

In this example, the homeowner needs to finance a **$20,500** off-grid solar system. They have multiple financing options, including personal savings, green energy loans, or leveraging government incentives. The choice depends on their financial situation, risk tolerance, and long-term plans. The investment not only provides energy independence but also potentially increases the property's value and offers long-term savings on electricity bills.

15.5 Example Calculation for ROI

- **Determine Initial Investment**

This includes the cost of solar panels, batteries, inverter, charge controller, installation, and any additional expenses such as permits or inspections.

Example: Total Initial Investment: $20,500

- **Calculate Annual Benefits**

Annual benefits are mainly the savings on electricity bills. Additionally, consider any incentives, rebates, or tax credits that have immediate financial benefits.

Example: Average Monthly Electricity Bill Without Solar: $150

Annual Savings: $150 x 12 = $1,800

- **Consider Ongoing Costs**

These include maintenance, potential repairs, or part replacements (especially batteries) over the system's lifespan.

Example: Average Annual Maintenance Cost: $200

- **Net Annual Savings**

Subtract the annual ongoing costs from the annual savings.

Example: Net Annual Savings: $1,800 (savings) - $200 (maintenance) = $1,600

- **Calculate Payback Period**

The payback period is the time it takes for the net savings to equal the initial investment.

Example:

Payback Period = Total Initial Investment / Net Annual Savings

Payback Period = $20,500 / $1,600 ≈ 12.8 years

- **ROI Calculation**

ROI is typically expressed as a percentage and can be calculated over a specific period, like the expected lifespan of the system.

Example:

Assume the lifespan of the system is 25 years.

Total Savings Over 25 Years = Net Annual Savings x 25

Total Savings Over 25 Years = $1,600 x 25 = $40,000

ROI = (Total Savings - Initial Investment) / Initial Investment

ROI = ($40,000 - $20,500) / $20,500

ROI ≈ 0.95 or 95%

- **ROI Interpretation**

An ROI of 95% over 25 years indicates a profitable investment.

It is also useful to compare this ROI with other potential investments to determine its relative profitability.

Conclusion

In this example, the off-grid solar system presents a substantial ROI of 95% over its 25-year lifespan, with a payback period of roughly 12.8 years. This analysis helps in understanding the long-term financial viability of the solar system investment. Remember, ROI can vary based on factors like energy usage patterns, system costs, and local energy prices.

Chapter 16
Off-Grid Solar Permit

Obtaining a permit for off-grid solar installations is an important step in ensuring your system is safe, compliant, and optimized for your needs. When planning a solar installation, it's crucial to be aware of the various types of permits that may be required, as these can vary based on your location, the type of solar system, and local regulations. Getting permits is often the longest step.

16.1 Permitting for Off-Grid Solar

If you are living in a rural area, there might not be any permitting requirements. This is good news because it saves installers time and money. However, if you are living in a city, there will likely be permitting requirements.

If you are unsure about the specific requirements in your area, the first step is to contact the local town or municipal authority. They can provide information on whether solar permits are required for off-grid photovoltaic (PV) systems. It's important to clarify this early in the planning stages to avoid any potential legal or administrative hurdles later on.

Solar permits typically require approval on two forms—one from your local building authority (to approve new construction), the other from your electric company (to approve interconnection to the grid). These are the common types of permits required for solar installations:

1. **Building Permits:**
    Building permits are essential for ensuring that the physical installation of solar panels, whether on a roof or as a ground mount, adheres to local building codes. This permit focuses on the structural aspects of the installation, including the stability of the roof or ground site, and ensures compliance with local zoning and building standards. The application typically requires a detailed structural analysis, especially for rooftop installations, to ensure that the roof can bear the additional weight of the solar panels. It also involves scrutiny of the mounting systems and overall installation plans to guarantee that they meet local safety standards.

2. **Electrical Permits:**
    Electrical permits are crucial for the safety and compliance of the electrical components of a solar installation. This includes wiring, inverters, junction

boxes, and the connection to the home's electrical system. These permits ensure that the installation adheres to the National Electrical Code (NEC) and local electrical codes. They require detailed electrical diagrams and specifications of all electrical components used. The primary focus is on safety, preventing electrical fires, and ensuring the system is capable of handling the electrical load safely.

3. **Utility Permits or Interconnection Agreements:**
   Since off-grid systems operate independently of the power grid, they do not require a utility interconnection permit, which is typically needed for grid-tied systems. Furthermore, off-grid systems do not qualify for net metering programs because they do not supply excess power back to the public grid.

4. **Fire Department Permits:**
   In some jurisdictions, fire department permits are required for solar installations, especially for those on rooftops. These permits are focused on ensuring that the installation does not impede firefighting operations and that there is adequate access and safety measures in place. The fire department reviews the layout of the solar panels to ensure that there is enough space for firefighters to walk on the roof and that the installation doesn't interfere with potential firefighting efforts.

5. **Environmental Permits:**
   Environmental permits are necessary in areas where solar installations might impact local ecosystems or protected environments. These permits assess the environmental impact of the installation, focusing on factors like land use, wildlife disturbance, and habitat conservation. They ensure that the installation of solar panels does not have a detrimental effect on the local environment and biodiversity.

16.2 Solar Permit Guide

- **Step 1: Research Local Requirements**

  Start by understanding the specific permitting requirements for your area, as these can vary significantly between rural and urban locations. Contact your local building and planning departments for detailed information. They will inform you about the types of permits necessary for an off-grid solar installation in your jurisdiction.

- **Step 2: Prepare Your Documentation**

Gather all the technical details of your solar system, including specifications of the solar panels, battery storage, inverter, and a comprehensive design of your system. You'll also need to create detailed site plans showing the exact location of your solar installation on your property, ensuring everything is clearly laid out.

- **Step 3: Building and Electrical Permits**

For the building permit, you'll need to submit an application along with your site plans and structural details of the installation. An electrical permit may also be necessary, requiring detailed electrical schematics of your solar setup to ensure adherence to local electrical codes.

- **Step 4: Environmental and Other Special Permits**

If your installation could potentially impact sensitive environmental areas, you might need to obtain relevant environmental permits. Also, if you're located in a historic district or an area with specific zoning regulations, additional permits or approvals could be required.

- **Step 5: Submitting Your Permit Applications**

Organize and compile all your applications, making sure they are complete with all necessary documentation. Then, submit these applications to the appropriate local authorities, either online or through physical submission, depending on the requirements.

- **Step 6: Review Process**

After submission, there will be a waiting period as your applications are reviewed. Be prepared for this to take some time and respond promptly to any additional information requests from the authorities.

- **Step 7: Getting Approval**

Once your applications are approved, you'll be notified, and you may receive physical permits. Some permits require post-installation inspections, so be sure to schedule these as needed.

- **Step 8: Installation and Final Inspection**

With your permits in hand, you can proceed with the solar installation. After completion, a final inspection by local authorities might be necessary to ensure that the installation meets all safety and code standards.

After passing the final inspection, your solar system can be officially commissioned. Remember to maintain your system according to best practices for efficient and safe operation.

This comprehensive guide should be helpful for your journey through the permitting process, ensuring that your DIY off-grid solar system is safe, legal, and ready for operation.

16.3 FAQs About Off-Grid Solar Permitting

1. **How Long Does Permitting Take?**
   Permitting for off-grid solar can vary from a few days to several weeks, depending on local office procedures and workload.

2. **What Are the Permitting Fees?**
   Permit fees usually range around a few hundred dollars but can vary based on location and project specifics.

3. **Do I Need to Handle Permitting Myself for a DIY Project?**
   If you are not confident, then the answer is No: you can use third-party services to manage the permitting process for a fee, offering convenience and expertise. These are two familiar 3rd party services for handling solar permitting processes. However, there are also other reputable services available in the market.

   - Gemini Solar Design
   - Solar Permit Services

4. **Is a Permit Required for Off-Grid Solar?**
   Yes, you generally need a permit for off-grid solar installations, despite not connecting to the grid. The primary reasons are to ensure fire safety and structural integrity. Local authorities need to confirm that your installation won't pose a risk to your property or its occupants and that it complies with local building codes. Even in off-grid scenarios, adhering to these regulations is crucial for a safe and reliable solar system.

Recommended Brands

There are several reputable solar product brands in the market. The best brand for you will depend on your specific needs, budget, and location. Here are some recommended brands, categorized by their products:

1. Solar Panels (PV Modules):

   - **SunPower**: Known for high efficiency and long warranties.

   - **LG**: Offers a range of high-quality solar panels.

   - **Panasonic**: Renowned for their efficiency and technological innovation.

   - **Canadian Solar**: Offers reliable modules at competitive prices.

   - **Trina Solar**: One of the world's largest solar panel manufacturers.

   - **JinkoSolar**: Widely used around the world and offers a range of panel options.

   - **Renogy**: Known for offering a range of solar panels suitable for both residential and portable applications.

2. Solar Inverters:

   - **SolarEdge**: Popular for their power optimizers and inverters.

   - **Enphase**: Known for their microinverters.

   - **SMA**: A leading manufacturer with a long-standing reputation.

   - **EPEVER**: Offers a variety of inverters suitable for different solar applications, known for their reliability and competitive pricing.

   - **Renogy**: Also offers inverters, particularly popular among off-grid and RV users.

   - **Fronius**: High-quality inverters especially popular in Europe and Australia.

3. **Solar Batteries:**

- **Tesla Powerwall**: One of the most recognized names in energy storage.

- **LG Chem**: Offers the RESU series which is popular in many markets.

- **Sonnen**: German-engineered batteries with a strong reputation.

- **SunPower**: SunPower is known for its high-quality solar panels, and their SunVault battery storage solution has earned a top spot for its performance and reliability.

- BYD: Chinese manufacturer known for their B-Box series of batteries.

4. **Solar Charge Controllers:**

- **EPEVER:** Renowned for their Tracer series of MPPT charge controllers. They offer reliable performance and are used in many off-grid systems around the world.

- **Victron Energy:** Offers a range of products, including charge controllers, known for quality and reliability.

- **Morningstar:** Highly regarded for their durable and efficient charge controllers, often chosen for off-grid and remote systems.

- **Renogy:** Offers a variety of charge controllers, especially suitable for smaller and portable systems.

- **OutBack Power**: Renowned for their durable and reliable charge controllers.

When selecting a brand, it's essential to:

- Check for warranties and what they cover.

- Review third-party testing results if available.

- Ask for local installer recommendations.

- Consider your location and the availability of after-sales service.

Remember that the solar industry is continuously evolving, with new products and companies emerging. Always do current research and seek out reviews or testimonials when making a purchasing decision.

Conclusion

As the final pages of "DIY Off-Grid Solar Power for Everyone" draw to a close, it's essential to reflect on the transformative journey we've embarked upon. We have delved deep into the heart of solar energy, dissecting its complexities, and emerged with a set of tools that empower even the most novice among us to harness the power of the sun.

Our exploration was not merely about understanding the principles of solar power. Instead, it was a hands-on guide, a mentor in written form, guiding you through the twists and turns of setting up your own solar-powered haven, be it a home, RV, van, or boat. We have championed the belief that with the right knowledge, everyone has the potential to be not just a consumer but a creator of clean, renewable energy.

While the step-by-step processes outlined within these pages are detailed and comprehensive, they represent more than just instructions. They symbolize a movement towards self-reliance, sustainability, and a commitment to a cleaner future. The DIY approach is not merely about saving costs; it's about understanding, appreciating, and having full control over where your energy comes from. It's about the satisfaction of setting up a panel, watching it drink in the sun, and knowing that you made it happen.

It's worth noting that the world of solar energy is ever-evolving. New advancements and refinements will continuously shape the landscape. Yet, the essence of this book remains timeless - empowering individuals to harness the abundant energy that our sun generously provides.

In closing, let this not be an end, but a beginning. A start of your adventures in the world of off-grid solar energy. Remember, every panel you install, every battery you charge, and every light you illuminate using solar energy is a step towards a sustainable future. And in this journey, always know that the power, quite literally, is in your hands.

Thank you for letting this guide be a part of your solar journey. Here's to a brighter, greener, and more empowered future for all!

This special page is exclusively for book owners, so remember to use the following password to gain access.

➢ Link: https://www.opengreenenergy.com/ebook01

➢ Password: **OGE_SOL@23**

Stay Connected!

For more interesting projects, insights, and updates, you can connect with me through the following platforms:

1. Blogpost: https://www.opengreenenergy.com

2. YouTube: https://www.youtube.com/c/opengreenenergy

3. Instagram: https://www.instagram.com/opengreenenergy

4. Instructables: https://www.instructables.com/member/opengreenenergy

I appreciate your continued support and interest in my work. Whether it's a quick question, a new project idea, or just a hello, I'd love to hear from you. If you have any more questions in the future, don't hesitate to reach out. Take care and best of luck with your endeavours in solar and beyond!

Thank You!

Debasish

www.ingramcontent.com/pod-product-compliance
Lightning Source LLC
Chambersburg PA
CBHW082130290526
45794CB00008B/2992